Also from Westphalia Press
westphaliapress.org

A History of Japanese Mathematics

by David Eugene Smith
and Yoshio Mikami

WESTPHALIA PRESS
An imprint of Policy Studies Organization

A History of Japanese Mathematics

Westphalia Press
An imprint of Policy Studies Organization
1527 New Hampshire Ave., NW
Washington, D.C. 20036
info@ipsonet.org

ISBN-13: 978-1-63391-852-8
ISBN-10: 1-63391-852-1

Cover design by Jeffrey Barnes:
jbarnesbook.design

Daniel Gutierrez-Sandoval, Executive Director
PSO and Westphalia Press

Updated material and comments on this edition
can be found at the Westphalia Press website:
www.westphaliapress.org

A HISTORY

OF

JAPANESE MATHEMATICS

BY

DAVID EUGENE SMITH

AND

YOSHIO MIKAMI

CHICAGO
THE OPEN COURT PUBLISHING COMPANY
1914

Printed by W. Drugulin, Leipzig

PREFACE

Although for nearly a century the greatest mathematical classics of India have been known to western scholars, and several of the more important works of the Arabs for even longer, the mathematics of China and Japan has been closed to all European and American students until very recently. Even now we have not a single translation of a Chinese treatise upon the subject, and it is only within the last dozen years that the contributions of the native Japanese school have become known in the West even by name. At the second International Congress of Mathematicians, held at Paris in 1900, Professor Fujisawa of the Imperial University of Tokio gave a brief address upon Mathematics of the old Japanese School, and this may be taken as the first contribution to the history of mathematics made by a native of that country in a European language. The next effort of this kind showed itself in occasional articles by Baron Kikuchi, as in the *Nieuw Archief voor Wiskunde*, some of which were based upon his contributions in Japanese to one of the scientific journals of Tokio. But the only serious attempt made up to the present time to present a well-ordered history of the subject in a European language is to be found in the very commendable papers by T. Hayashi, of the Imperial University at Sendai. The most important of these have appeared in the *Nieuw Archief voor Wiskunde*, and to them the authors are much indebted.

Having made an extensive collection of mathematical manuscripts, early printed works, and early instruments, and having brought together most of the European literature upon the subject and embodied it in a series of lectures for my classes in the history of mathematics, I welcomed the suggestion of Dr. Carus that I join with Mr. Mikami in the preparation of the present work. Mr. Mikami has already made for himself an enviable reputation as an authority upon the *wasan* or native Japanese mathematics, and his contributions to the *Bibliotheca Mathematica* have attracted the attention of western scholars. He has also published, as a volume of the *Abhandlungen zur Geschichte der Mathematik*, a work entitled *Mathematical Papers from the Far East.* Moreover his labors with the learned T. Endō, the greatest of the historians of Japanese mathematics, and his consequent familiarity with the classics of his country, eminently fit him for a work of this nature.

Our labors have been divided in the manner that the circumstances would suggest. For the European literature, the general planning of the work, and the final writing of the text, the responsibility has naturally fallen to a considerable extent upon me. For the furnishing of the Japanese material, the initial translations, the scholarly search through the excellent library of the Academy of Sciences of Tokio, where Mr. Endō is librarian, and the further examination of the large amount of native secondary material, the responsibility has been Mr. Mikami's. To his scholarship and indefatigable labors I am indebted for more material than could be used in this work, and whatever praise our efforts may merit should be awarded in large measure to him.

The aim in writing this work has been to give a brief survey of the leading features in the development of the *wasan.* It has not seemed best to enter very fully into the details of demonstration or into the methods of solution employed by the great writers whose works are described. This would not be done in a general history of European mathematics, and there is no reason why it should be done here, save in cases where some peculiar feature is under discussion. Undoubtedly several names of importance have been omitted, and at least a score of names that might properly have had mention have been the subject of correspondence between the authors for the past year. But on the whole it may be said that most of those writers in whose works European scholars are likely to have much interest have been mentioned.

It is the hope of the authors that this brief history may serve to show to the West the nature of the mathematics that was indigenous to Japan, and to strengthen the bonds that unite the scholars of the world through an increase in knowledge of and respect for the scientific attainments of a people whose progress in the past four centuries has been one of the marvels of history.

It is only just to mention at this time the generous assistance rendered by Mr. Leslie Leland Locke, one of my graduate students in the history of mathematics, who made in my library the photographs for all of the illustrations used in this work. His intelligent and painstaking efforts to carry out the wishes of the authors have resulted in a series of illustrations that not merely elucidate the text, but give a visual idea of the genius of the Japanese mathematics that words alone cannot give. To him I take pleasure in ascribing the credit for this arduous labor, and in expressing the thanks of the authors.

Teachers College,
Columbia University,
New York City,
December 1, 1913.

David Eugene Smith.

VOCABULARY FOR REFERENCE

The following brief vocabulary will be convenient for reference in considering some of the Japanese titles:

hō, method or theory. Synonym of *jutsu*. It is found in expressions like *shōsa hō* (method of differences).

hyō, table.

jutsu, method or theory. Synonym of *hō*, It is found in words like *kaku jutsu* (polygonal theory) and *tatsujutsu* (method of expanding a root of a literal equation).

ki, a treatise.

roku, a treatise. Synonym of *ki*.

sampō, mathematical treatise, or mathematical rules.

sangi, rods used in computing, and as numerical coefficients in equations

soroban, the Japanese abacus.

tengen, celestial element. The Japanese name for the Chinese algebra.

tenzan, the algebra of the Seki school.

wasan, the native Japanese mathematics as distinguished from the *yōsan*, the European mathematics.

yenri, circle principle. A term applied to the native calculus of Japan.

In Japanese proper names the surname is placed first in accordance with the native custom, excepting in the cases of persons now living and who follow the European custom of placing the surname last.

CONTENTS

CHAPTER I.

The Earliest Period.

The history of Japanese mathematics, from the most remote times to the present, may be divided into six fairly distinct periods. Of these the first extended from the earliest ages to 552[1], a period that was influenced only indirectly if at all by Chinese mathematics. The second period of approximately a thousand years (552--1600) was characterized by the influx of Chinese learning, first through Korea and then direct from China itself, by some resulting native development, and by a season of stagnation comparable to the Dark Ages of Europe. The third period was less than a century in duration, extending from about 1600 to the beginning of Seki's influence (about 1675). This may be called the Renaissance period of Japanese mathematics, since it saw a new and vigorous importation of Chinese science, the revival of native interest through the efforts of the immediate predecessors of Seki, and some slight introduction of European learning through the early Dutch traders and through the Jesuits. The fourth period, also about a century in length (1675 to 1775) may be compared to the synchronous period in Europe. Just as the initiative of Descartes, Newton, and Leibnitz prepared the way for the labors of the Bernoullis, Euler, Laplace, D'Alembert, and their contemporaries of the eighteenth century, so the work of the great Japanese teacher, Seki, and of his pupil Takebe, made possible a noteworthy development of the *wasan*[2] of Japan during the same

---

[1] All dates are expressed according to the Christian calendar and are to be taken as after Christ unless the contrary is stated.

[2] The native mathematics, from *Wa* (Japan) and *san* (mathematics). The word is modern, having been employed to distinguish the native theory from the western mathematics, the *yōsan*.

century. The fifth period, which might indeed be joined with the fourth, but which differs from it much as the nineteenth century of European mathematics differs from the eighteenth, extended from 1775 to 1868, the date of the opening of Japan to the Western World. This is the period of the culmination of native Japanese mathematics, as influenced more or less by the European learning that managed to find some entrance through the Dutch trading station at Nagasaki and through the first Christian missionaries. The sixth and final period begins with the opening of Japan to intercourse with other countries and extends to the present time, a period of marvelous change in government, in ideals, in art, in industry, in education, in mathematics and the sciences generally, and in all that makes a nation great. With these stupendous changes of the present, that have led Japan to assume her place among the powers of the world, there has necessarily come both loss and gain. Just as the world regrets the apparent submerging of the exquisite native art of Japan in the rising tide of commercialism, so the student of the history of mathematics must view with sorrow the necessary decay of the *wasan* and the reduction or the elevation of this noble science to the general cosmopolitan level. The mathematics of the present in Japan is a broader science than that of the past; but it is no longer Japanese mathematics,—it is the mathematics of the world.

It is now proposed to speak of the first period, extending from the most remote times to 552. From the nature of the case, however, little exact information can be expected of this period. It is like seeking for the early history of England from native sources, excluding all information transmitted through Roman writers. Egypt developed a literature in very remote times, and recorded it upon her monuments and upon papyrus rolls, and Babylon wrote her records upon both stone and clay; but Japan had no early literature, and if she possessed any ancient written records they have long since perished.

It was not until the fifteenth year of the Emperor Ōjin (284), so the story goes, that Chinese ideograms, making their way through Korea, were first introduced into Japan. Japanese nobles now began to learn to read and write, a task of enormous difficulty in the Chinese system. But the records themselves have long since perished, and if they contained any knowledge of mathematics, or if any mathematics from China at that time reached the shores of Japan, all knowledge of this fact has probably gone forever. Nevertheless there is always preserved in the language of a people a great amount of historical material, and from this and from folklore and tradition we can usually derive some little knowledge of the early life and customs and number-science of any nation.

So it is with Japan. There seems to have been a number mysticism there as in all other countries. There was the usual reaching out after the unknown in the study of the stars, of the elements, and of the essence of life and the meaning of death. The general expression of wonder that comes from the study of number, of forms, and of the arrangements of words and objects, is indicated in the language and the traditions of Japan as in the language and traditions of all other peoples. Thus we know that the *Jindai monji*, "letters of the era of the gods",[1] go back to remote times, and this suggests an early cabala, very likely with its usual accompaniment of number values to the letters; but of positive evidence of this fact we have none, and we are forced to rely at present only upon conjecture.[2]

Practically only one definite piece of information has come

---

[1] Nothing definite is known as to these letters. They may have been different alphabetic forms. *Monji* (or *moji*) means letters, *Jin* is god, and *dai* is the age or era. The expression may also be rendered "letters of the age of heros", using the term hero to mean a mythological semi-divinity, as it is used in early Greek lore.

[2] There is here, howewer, an excellent field for some Japanese scholar to search the native folklore for new material. Our present knowledge of the *Jindai* comes chiefly from a chapter in the *Nihon-gi* (Records of Japan) entitled *Jindai no Maki* (Records of the Gods' Age), written by Prince Toneri Shinnō in 720. This is probably based upon early legends handed down by the *Kataribe*, a class of men who in ancient times transmitted the legends orally, somewhat like the old English bards.

down to us concerning the very early mathematics of Japan, and this relates to the number system. Tradition tells us that in the reign of Izanagi-no-Mikoto, the ancestor of the Mikados, long before the unbroken dynasty was founded by Jimmu (660 B. C.), a system of numeration was known that extended to very high powers of ten, and that embodied essentially the exponential law used by Archimedes in his *Sand Reckoner*[1] that

$$a^m a^n = a^{m+n}.$$

In this system the number names were not those of the present, but the system may have been the same, although modern Japanese anthropologists have serious doubts upon this matter. The following table[2] has been given as representing the ancient system, and it is inserted as a possibility, but the whole matter is in need of further investigation:

	Ancient	*Modern*		*Ancient*	*Modern*
1	hito	ichi	100	momo	hyaku
2	futa	ni	1000	chi	sen
3	mi	san	10000	yorozu	man
4	yo	shi	100000	so yorozu	jiu man
5	itsu	go	1000000	momo yorozu	hyaku man
6	mu	roku	10000000	chi yorozu	sen man
7	nana	shichi	100000000	yorozu yorozu	oku
8	ya	hachi	1000000000	so yorozu yorozu	jiu oku
9	koko	ku			
10	tō	jiu			

1 Ψαμμίτης, *De harena numero*, as it appears in Basel edition of 1544.

2 Endō, T., *Dai Nihon Sūgaku Shi* (*History of Japanese mathematics*, in Japanese. Tokio 1896, Book I, pp. 3—5; hereafter referred to as Endō). See also Knott, C. G., *The Abacus in its historic and scientific aspects*, in the *Transactions of the Asiatic Society of Japan*, Yokohama 1886, vol. XIV, p. 38; hereafter referred to as Knott. Another interesting form of counting is still in use in Japan, and is more closely connected with the ancient one than is the common form above given. It is as follows: (1) hitotsu, (2) futatsu, (3) mittsu, (4) yottsu, (5) itsutsu, (6) muttsu, (7) nanatsu, (8) yattsu, (9) kokonotsu, (10) tō. Still another form at present in use, and also related to the ancient one, is as follows: (1) hi, (2) fu, (3) mi, (4) yō, (5) itsu, (6) mū, (7) nana, (8) ya, (9) kono, (10) tō. Each of these forms is used only in counting, not in naming numbers, and their persistence may be compared

The interesting features of the ancient system are the decimal system and the use of the word *yorozu*, which now means 10000. This, however, may be a meaning that came with the influx of Chinese learning, and we are not at all certain that in ancient Japanese it stood for the Greek myriad.[1] The use of *yorozu* for 10000 was adopted in later times when the number names came to be based upon Chinese roots, and it may possibly have preceded the entry of Chinese learning in historic times. Thus 10^5 was not "hundred thousand"[2] in this later period, but "ten myriads",[3] and our million[4] is a hundred myriads.[5] Now this system of numeration by myriads is one of the frequently observed evidences of early intercourse between the scholars of the East and the West. Trades people and the populace at large did not need such large numbers, but to the scholar they were significant. When, therefore, we find the myriad as the base of the Greek system,[6] and find it more or less in use in India,[7] and know that it still persists in China,[8] and see it systematically used in the ancient Japanese system as well as in the modern number names, we are

---

with that of the "counting out" rhymes of Europe and America. It should be added that the modern forms given above are from Chinese roots.

1 Μυρίοι, 10000.

2 Which would, if so considered, appear as *momo chi*, or in modern Japanese as *hyaku sen*.

3 *So yorozu*, a softened form of *tō yorozu*. In modern Japanese, *jiu man*, *man* being the myriad.

4 Mille + on, "big thousand", just as saloon is salle + on, a big hall, and gallon is gill + on, a big gill.

5 *Momo yorozu*, or, in modern Japanese, *hyaku man*.

6 See, for example, Gow, J., *History of Greek Mathematics*, Cambridge 1884, and similar works.

7 See Colebrooke, H. T., *Algebra, with Arithmetic and Mensuration, from the Sanscrit of Brahmegupta and Bhascara*. London 1817, p. 4; Taylor, J., *Lilawati*. Bombay 1816, p. 5.

8 Williams, S. W., *The Middle Kingdom*. New York 1882; edition of 1895, vol. I, p. 619. Thus *Wan sui* is a myriad of years, and *Wan sui Yeh* means the Lord of a Myriad Years, *i. e.*, the Emperor. The swastika is used by the Buddhists in China as a symbol for myriad. This use of the myriad in China is very ancient.

convinced that there must have been a considerable intercourse of scholars at an early date.[1]

Of the rest of Japanese mathematics in this early period we are wholly ignorant, save that we know a little of the ancient system of measures and that a calendar existed. How the merchants computed, whether the almost universal finger computation of ancient peoples had found its way so far to the East, what was known in the way of mensuration, how much of a crude primitive observation of the movements of the stars was carried on, what part was played by the priest in the orientation of shrines and temples, what was the mystic significance of certain numbers, what, if anything, was done in the recording of numbers by knotted cords, or in representing them by symbols,—all these things are looked for in the study of any primitive mathematics, but they are looked for in vain in the evidences thus far at hand with respect to the earliest period of Japanese history. It is to be hoped that the spirit of investigation that is now so manifest in Japan will result in throwing more light upon this interesting period in which mathematics took its first root upon Japanese soil.

---

[1] There is considerable literature upon this subject, and it deserves even more attention. See, for example, the following: KLINGSMILL, T. W., *The Intercourse of China with Eastern Turkestan ... in the second century B. C.*, in the *Journal of the Royal Asiatic Society*, N. S., London 1882, vol. XIV, p. 74. A Japanese scholar, T. Kimura, is just at present maintaining that his people have a common ancestry with the races of the Greco-Roman civilization, basing his belief upon a comparison of the mythology and the language of the two civilizations. See also P. VON BOHLEN, *Das alte Indien mit besonderer Rücksicht auf Ægypten.* Königsberg 1830; REINAUD, *Relations politiques et commerciales de l'Empire Romain avec l'Asie orientale.* Paris 1863; P. A. DI SAN FILIPO, *Delle Relazioni antiche et moderne fra L'Italia e l'India.* Rome 1886; SMITH and KARPINSKI, *The Hindu-Arabic Numerals.* Boston 1911, with extensive bibliography on this point.

CHAPTER II.

The Second Period.

The second period in the history of Japanese mathematics (552—1600) corresponds both in time and in nature with the Dark Ages of Europe. Just as the Northern European lands came in contact with the South, and imbibed some slight draught of classical learning, and then lapsed into a state of indifference except for the influence of an occasional great soul like that of Charlemagne or of certain noble minds in the Church, so Japan, subject to the same *Zeitgeist*, drank lightly at the Chinese fountain and then lapsed again into semi-barbarism. Europe had her Gerbert, and Leonardo of Pisa, and Sacrobosco, but they seem like isolated beacons in the darkness of the Middle Ages; and in the same way Japan, as we shall see, had a few scholars who tended the lamp of learning in the medieval night, and who are known for their fidelity rather than for their genius.

Just as in the West we take the fall of Rome (476) and the fall of Constantinople (1453), two momentous events, as convenient limits for the Dark Ages, so in Japan we may take the introduction of Buddhism (552) and the revival of learning (about 1600) as similar limits, at least in our study of the mathematics of the country.

It was in round numbers a thousand years after the death of Buddha[1] that his religion found its way into Japan.[2] The

1 The Shinshiu or "True Sect" of Buddhists place his death as early as 949 B. C., but the Singalese Buddhists place it at 543 B. C. Rhys Davids, who has done so much to make Buddhism known to English readers, gives 412 B. C., and Max Müller makes it 477 B. C., See also SUMNER, J., *Buddhism and traditions concerning its introduction into Japan, Transactions of the Asiatic Society of Japan,* Yokohama 1886, vol. XIV, p. 73. He gives the death of Buddha as 544 B. C.

2 It was introduced into China in 64 A. D., and into Korea in 372.

date usually assigned to this introduction is 552, when an image of Buddha was set up in the court of the Mikado; but evidence[1] has been found which leads to the belief that in the sixteenth year of Keitai Tenno (an emperor who reigned in Japan from 507 to 531), that is in the year 522, a certain man named Szŭ-ma Ta[2] came from Nan-Liang[3] in China, and set up a shrine in the province of Yamato, and in it placed an image of Buddha, and began to expound his religion. Be this as it may, Buddhism secured a foothold in Japan not far from the traditional date of 552, and two years later[4] Wang Pao-san, a master of the calendar,[5] and Wang Pao-liang, doctor of chronology,[6] an astrologer, crossed over from Korea and made known the Chinese chronological system. A little later a Korean priest named Kanroku[7] crossed from his native country and presented to the Empress Suiko a set of books upon astrology and the calendar.[8] In the twelfth year of her reign (604) almanacs were first used in Japan, and at this period Prince Shōtoku Taishi proved himself such a fosterer of Buddhism and of learning that his memory is still held in high esteem. Indeed, so great was the fame of Shōtoku Taishi that tradition makes him the father of Japanese arithmetic and even the inventor of the abacus.[9] (Fig. 1.)

A little later the Chinese system of measures was adopted, and in general the influence of China seems at once to have

---

1 See SUMNER, loc. cit., p. 78.

2 In Japanese, Shiba Tatsu.

3 I. e., South Liang, Liang being one of the southern monarchies.

4 I. e., in 554, or possibly 553.

5 In Europe he would have had charge of the Compotus, the science of the Church calendar, in a Western monastery.

6 Also called a Doctor of Yih. The doctrine of Yih (changes) is set forth in the *Yih King* (Book of Changes), one of the ancient Five Classics of the Chinese. There is a very extensive literature upon this subject.

7 Or Ch'üan-lo.

8 SUMNER, loc. cit, p. 80, gives the date as 593. Endō, who is the leading Japanese authority, gives it as 602.

9 That this is without foundation will appear in Chapter III. The *soroban* which he holds in the illustration here given is an anachronism.

become very marked. Fortunately, just about this time, the Emperor Tenchi (Tenji) began his short but noteworthy reign (668—671).[1] While yet crown prince this liberal-minded man invented a water clock, and divided the day into a hundred hours, and upon ascending the throne he showed his further interest by founding a school to which two doctors of arithmetic and twenty students of the subject were appointed. An observatory was also established, and from this time mathematics had recognized standing in Japan.

Fig. 1.
Shōtoku Taishi, with a *soroban*. From a bronze statuette.

The official records show that a university system was established by the Emperor Monbu in 701, and that mathematical studies were recognized and were regulated in the higher institutions of learning. Nine Chinese works were specified, as follows:—(1) *Chou-pei* (*Suan-ching*), (2) *Sun-tsu* (*Suan-ching*), (3) *Liu-chang*, (4) *San-k'ai Chung-ch'a*, (5) *Wu-t'sao* (*Suan-shu*), (6) *Hai-tao* (*Suan-shu*), (7) *Chiu-szu*, (8) *Chiu-chang*, (9) *Chui-shu*.[2] Of these works, apparently the most famous of their time, the third, fourth, and seventh are lost. The others are probably known, and although they are not of native Japanese production they so greatly influenced the mathematics of Japan as to deserve some description at this time. We shall therefore consider them in the order above given.

1. *Chou-pei Suan-ching*. This is one of the oldest of the Chinese works on mathematics, and is commonly known in

---

[1] MURRAY, D., *The Story of Japan*. N. Y. 1894, p. 398, from the official records.

[2] ENDŌ, Book I, pp. 12—13.

China as *Chow-pi,* said to mean the "Thigh bone of Chow".[1] The thigh bone possibly signifies, from its shape, the base and altitude of a triangle. Chow is thought to be the name of a certain scholar who died in 1105 B. C., but it may have been simply the name of the dynasty. This scholar is sometimes spoken of as Chow Kung,[2] and is said to have had a discussion with a nobleman named Kaou, or Shang Kao,[3] which is set forth in this book in the form of a dialogue. The topic is our so-called Pythagorean theorem, and the time is over five hundred years before Pythagoras gave what was probably the first scientific proof of the proposition. The work relates to geometric measures and to astronomy.[4]

2. *Sun-tsu Suan-ching.* This treatise consists of three books, and is commonly known in China as the *Swan-king* (Arithmetical classic) of Sun-tsu (Sun-tsze, or Swen-tse), a writer who lived probably in the 3d century A. D., but possibly much earlier. The work attracted much attention and is referred to by most of the later writers, and several commentaries have appeared upon it. Sun-tsu treats of algebraic quantities, and gives an example in indeterminate equations. This problem is to "find a number which, when divided by 3 leaves a remainder of 2, when divided by 5 leaves 3, and when divided by 7 leaves 2."[5] This work is sometimes, but without any good reason, assigned to Sun Wu, one of the most illustratious men of the 6th century B. C.

3. *Liu-Chang.* This is unknown. There was a writer named

---

1 *Pi* means leg, thigh, thigh-bone.

2 Chi Tan, known as Chow Kung (that is, the Duke of Chow), was brother and advisor to the Emperor Wu Wang of the Chow dynasty. It is possible that he wrote the Chow Li, "Institutions of the Chow Dynasty", although it is more probable that it was written for him. The establishment and prosperity of the Chow dynasty is largely due to him. There is no little doubt as to the antiquity of this work, and the critical study of scholars may eventually place it much later than the traditional date here given.

3 Also written Shang Kaou.

4 For a translation of the dialogue see WYLIE, A., *Chinese Researches.* Shanghai 1897, Part III, p. 163.

5 His result is 23. For his method of solving see WYLIE, loc. cit., p. 175.

Liu Hui[1] who wrote a treatise entitled *Chung-ch'a*, but this seems to be No. 4 in the list.

4. *San-k'ai Chung-ch'a*. This is also unknown, but is perhaps Liu Hui's *Chung-ch'a-keal-tsih-wang-chi-shuh* (The whole system of measuring by the observation of several beacons), published in 263. The author also wrote a commentary on the *Chiu-chang* (No. 8 in this list). It relates to the mensuration of heights and distances, and gives only the rules without any explanation. About 1250 Yang Hway published a work entitled *Siang-kiai-Kew-chang-Swan-fa* (Explanation of the arithmetic of the Nine Sections), but this is too late for our purposes. He also wrote a work with a similar title *Siang-kiai-Jeh-yung-Swan-fa* (Explanation of arithmetic for daily use).

5. *Wu-t'sao Suan-shu*. The author and the date of this work are both unknown, but it seems to have been written in the 2d or 3d century.[2] It is one of the standard treatises on arithmetic of the Chinese.

6. *Hai-tao Suan-shu*. This was a republication of No. 4, and appeared about the time of the Japanese decree of 701. The name signifies "The Island Arithmetical Classic",[3] and seems to come from the first problem, which relates to the measuring of an island from a distant point.

7. *Chiu-szu*. This work, which was probably a commentary on the *Suan-shu* (*Swan-king*) of No. 8, is lost.

8. *Chiu-chang*. *Chiu-chang Suan-shu*[4] means "Arithmetical Rules in Nine Sections". It is the greatest arithmetical classic of China, and tradition assigns to it remote antiquity. It is related in the ancient *Tung-kien-kang-muh* (General History of China) that the Emperor Hwang-ti,[5] who lived in 2637 B. C.,

---

1 Lew-hwuy according to Wylie's transliteration, who also assigns him to about the 5th century B. C.

2 But see Wylie, loc. cit., who refers it to about the 5th century, and improperly states that Wu-t'sao is the author's name. He gives it the common name of *Swan-king* (Arithmetical classic).

3 Also written *Hae-taou-swan-king*.

4 *Kew-chang-swan-shu*, *Kiu-chang-san-suh*, *Kieou chang*.

5 Or Hoan-ti, the "Yellow Emperor". Some writers give the date much earlier.

caused his minister Li Show[1] to form the *Chiu-chang.*[2] Of the text of the original work we are not certain, for the reason that during the Ch'in dynasty (220—205 B. C.) the emperor Chi Hoang-ti decreed, in 213 B. C., that all the books in the empire should be burned. And while it is probable that the classics were all surreptitiously preserved, and while they could all have been repeated from memory, still the text may have been more or less corrupted during the reign of this oriental vandal. The text as it comes to us is that of Chang T'sang of the second century B. C., revised by Ching Ch'ou-ch'ang about a hundred years later.[3] Both of these writers lived in the Former Han[4] dynasty (202 B. C.—24 A. D.), a period corresponding in time and in fact with the Augustan age in Europe, and one in which great effort was made to restore the lost classics,[5] and both were ministers of the emperor.

This classical work had such an effect upon the mathematics of Japan that a summary of the contents of the books or chapters of which it is composed will not be out of place. The work contained 246 problems, and these are arranged in nine sections as follows:

(1) *Fang-tien,* surveying. This relates to the mensuration of various plane figures, including triangles, quadrilaterals, circles, circular segments and sectors, and the annulus. It also contains some treatment of fractions.

(2) *Suh-pu* (*Shu-poo*). This treats chiefly of commercial problems solved by the "rule of three".

(3) *Shwai-fên* (*Shwae-fun, Shuai-fen*). This deals with partnership.

---

[1] Or Li-shou.

[2] WYLIE, A., *Jottings of the Science of Chinese Arithmetic, North China Herald* for 1852, *Shanghai Almanac* for 1853, *Chinese Researches,* Shanghai 1897, Part III, page 159; BIERNATZKI, *Die Arithmetik der Chinesen,* CRELLE's *Journal* for 1856, vol. 52.

[3] For this information we are indebted to the testimony of Liu Hui, whose commentary was written in 263.

[4] Also known as the Western Han.

[5] LEGGE, J., *The Chinese Classics.* Oxford 1893, 2nd edition, vol. I, p. 4.

(4) *Shao-kang* (*Shaou-kwang*). This relates to the extraction of square and cube roots, the process being much like that of the present time.

(5) *Shang-kung*. This has reference to the mensuration of such solids as the prism, cylinder, pyramid, circular cone, frustum of a cone, tetrahedron, and wedge.

(6) *Kin-shu* (*Kiun-shoo*, *Ghün-shu*) treats of alligation.

(7) *Ying-pu-tsu* (*Yung-yu*, *Yin-nuh*). This chapter treats of "Excess and deficiency", and follows essentially the old rule of false position.[1]

(8) *Fang-ch'êng* (*Fang-chêng*, *Fang-ching*). This chapter relates to linear equations involving two or more unknown quantities, in which both positive (ching) or negative (foo) terms are employed. The following example is a type: "If 5 oxen and 2 sheep cost 10 taels of gold, and 2 oxen and 8 sheep cost 8 taels, what is the price of each?" It is probable that this chapter contains the earliest known mention of a negative quantity, and if the ancient text has not been corrupted, it places this kind of number between 2000 and 3000 B. C.

(9) *Kou-ku*, a term meaning a right triangle. The essential feature of this chapter is the Pythagorean theorem, which is stated as follows: "The first side and the second side being each squared and added, the square root of the sum is the hypotenuse." One of the twenty-four problems in this section involves the equation $x^2 + (20 + 14)x - 2 \times 20 \times 1775 = 0$, and a rule is laid down that is equivalent to the modern formula for the quadratic. If these problems were in the original text, and that text has the antiquity usually assigned to it, concerning neither of which we are at all certain, then they contain the oldest known quadratic equation. The interrelation of ancient mathematics is seen in two problems in this chapter. One is that of the reed growing 1 foot above the surface in the center of a pond 10 feet square, which just reaches the surface when drawn to the edge of the pond, it being required to find the

[1] The *Regula falsi* or *Regula positionis* of the Middle Ages in Europe. The rule seems to have been of oriental origin.

depth of the water. The other is the problem of the broken tree that has been a stock question for four thousand years. Both of these problems are found in the early Hindu works and were among the medieval importations into Europe.

The value of π[1] used in the "Nine Sections" is 3, as was the case generally in early times.[2] Commentators changed this later, Liu Hui (263) giving the value $\frac{157}{50}$, which is equivalent to 3.14.[3]

9. *Chui-shu.* This is usually supposed to be Tsu Ch'ung-chih's work which has been lost and is now known only by name.

This list includes all of the important Chinese classics in mathematics that had appeared before it was made, and it shows a serious attempt to introduce the best material available into the schools of Japan at the opening of the 8th century. It seemed that the country had entered upon an era of great intellectual prosperity, but it was like the period of Charlemagne, so nearly synchronous with it,—a temporary beacon in a dark night. Instead of leading scholars to the study of pure mathematics, this introduction of Chinese science, at a time when the people were not fully capable of appreciating it, seemed rather to foster a study of astrology, and mathematics degenerated into mere puzzle solving, the telling of fortunes, and the casting of horoscopes. Japan itself was given up to wars and rumors of wars. The "Nine Sections" was forgotten, and a man who actually knew arithmetic was looked upon as a genius. The *samurai* or noble class disdained all commercial pursuits, and ability to operate with numbers was looked upon as evidence of low birth. Professor Nitobe has given us a picture of this feudal society in his charming little book entitled *Bushido, The Soul of Japan.*[4] "Children," he

---

1 In Chinese *Chou-le*; in Japanese *yenshū ritsu.*

2 It is also found in the *Chou-pei*, No. 1 in this list.

3 MIKAMI, Y., *On Chinese Circle-Squarers*, in the *Bibliotheca Mathematica*, 1910, vol. X (3), p. 193.

4 Tokio 1905, p. 88. Some historical view of these early times is given in an excellent work by W. H. SHARP, *The Educational System of Japan.* Bombay 1906, pp. 1, 10, 11.

says, "were brought up with utter disregard of economy. It was considered bad taste to speak of it, and ignorance of the value of different coins was a token of good breeding. Knowledge of numbers was indispensable in the mastering of forces as well as in the distribution of benefices and fiefs, but the counting of money was left to meaner hands." Only in the Buddhist temples in Japan, as in the Christian church schools in Europe, was the lamp of learning kept burning.[1] In each case, however, mathematics was not a subject that appealed to the religious body. A crude theology, a purposeless logic, a feeble literature,—these had some standing; but mathematics save for calendar purposes was ever an outcast in the temple and the church, save as it occasionally found some eccentric individual to befriend it. In the period of the Ashikaga shoguns it is asserted that there hardly could be found in all Japan a man who was versed in the art of division.[2] To divide, the merchant resorted to the process known as *Shokei-zan*, a scheme of multiplication[3] which seems in some way to have served for the inverse process as well.[4] Nevertheless the assertion that the art of division was lost during this era of constant wars is not exact. Manuscripts on the calendar, corresponding to the European compotus rolls, and belonging to the period in question, contain examples of division, and it is probable that here, as in the West, the religious communities always had someone who knew the rudiments of calendar-reckoning. (Fig. 2.)

Three names stand out during these Dark Ages as worthy of mention. The first is that of Tenjin, or Michizane, counsellor and teacher in the court of the Emperor Uda (888—898).

---

1 Notably in the case of the labors of the learned Kōbō Daishi, founder of the Chĕnyen sect of Buddhists, who was born in 774 A. D. See Professor T. TANIMOTO's address on Kōbō Daishi. Kobe 1907.

2 ENDŌ, Book I, p. 30.

3 UCHIDA GOKAN, *Kokon Sankwan*, 1832, preface.

4 This is the opinion of MURAI CHŪZEN who lived in the 18th century. See his *Sampō Dōshi-mon*. 1781. Book I, article on the origin of arithmetic.

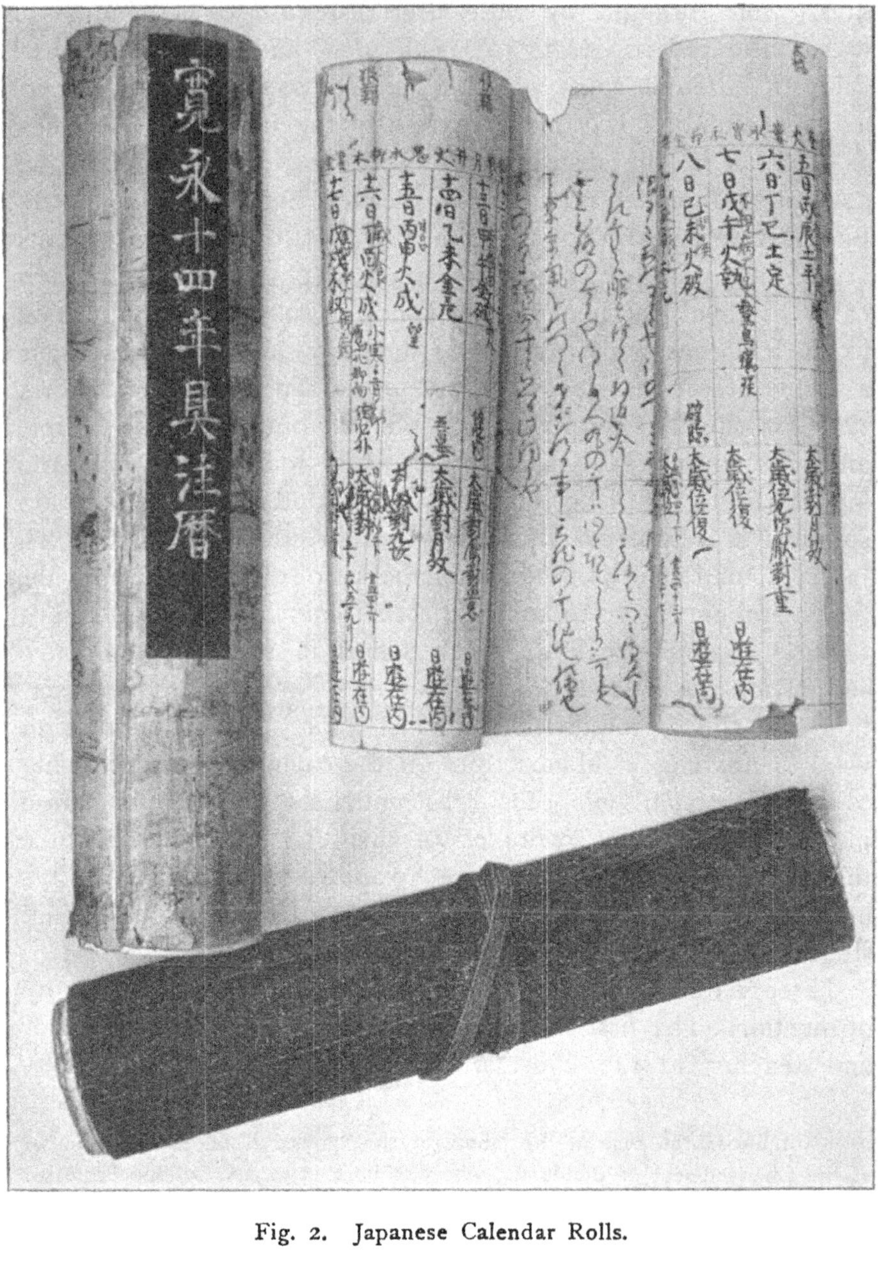

Fig. 2. Japanese Calendar Rolls.

Uda's successor, Daigo, banished him from the court and he died in 903. He was a learned man, and after his death he was canonized under the name Tenjin (Heavenly man) and was looked upon as the patron of science and letters. (See Fig. 3.) The second is that of Michinori, Lord of the province of Hyūga. His name is connected with a mathematical theory called the *Keishi-zan.*[1] It seems to have been related to permutations and to have been thought of enough consequence to attract the attention of Yoshida[2] and of his great successor Seki[3] in the 17th century. Michinori's work was written in the Hogen period (1156—1159).

Fig. 3. Tenjin, from an old bronze.

The third name is that of Genshō, a Buddhist priest in the time of Shogun Yoriyiye, at the opening of the 13th century. Tradition[4] says that he was distinguished for his arithmetical powers, but so far as we know he wrote nothing and had no permanent influence upon mathematics.

Thus passes and closes a period of a thousand years, with not a single book of any merit, and without advancing the science of mathematics a single pace. Europe was backward enough, but Japan was worse. China was doing a little, India was doing more, but the Arab was accomplishing still more through his restlessness of spirit if not through his mathematical genius. The world's rebirth was approaching, and this Renaissance came to Japan at about the time that it came to Europe, accompanied in both cases by a grafting of foreign learning upon native stock.

---

[1] ENDŌ, Book I, p. 28; Murai Chuzen, *Sampō Dōshimon.*

[2] See his *Jinkō-ki* of 1627.

[3] See Chapter VI.

[4] See ISOMURA KITTOKU, *Shusho Ketsugishō,* 1684, Book 4, marginal note. Isomura died in 1710.

CHAPTER III.

The Development of the Soroban.

Before proceeding to a consideration of the third period of Japanese mathematics, approximately the seventeenth century of the Christian era, it becomes necessary to turn our attention to the history of the simple but remarkable calculating machine which is universal in all parts of the Island Empire, the *soroban*. This will be followed by a chapter upon another mechanical aid known as the *sangi*, since each of these devices had a marked influence upon higher as well as elementary mathematics from the seventeenth to the nineteenth century.[1]

The numeral systems of the ancients were so unsuited to the purposes of actual calculation that probably some form of mechanical calculation was always necessary. This fact is the more evident when we consider that convenient writing material

---

[1] The literature of these forms of the abacus is extensive. The following are some of the most important sources: VISSIÈRE, A., *Recherches sur l'origine de l'abacque chinois*, in *Bulletin de Géographie*. Paris 1892; KNOTT, C. G., *The Abacus in its historic and scientific aspects*, in the *Transactions of the Asiatic Society of Japan*, Yokohama 1886, vol. 14, p. 18; GOSCHKEWITSCH, J., *Ueber das Chinesiche Rechenbrett*, in the *Arbeiten der Kaiserlich Russischen Gesandschaft zu Peking*, Berlin 1858, vol. I, p. 293 (no history); VAN NAME, R., *On the Abacus of China and Japan*, *Journal of the American Oriental Society*, 1875, vol. X, proc., p. CX; RODET, L., *Le souan-pan des Chinois*, *Bulletin de la Societé mathématique de France*, 1880, vol. VIII; DE LA COUPERIE, A. T., *The Old Numerals, the Counting-Rods, and the Swan-pan*, *Numismatic Chronicle*, London 1883, vol. III (3), p. 297; HAYASHI, T., *A brief history of Japanese Mathematics*, part I, p. 18; HÜBNER, M., *Die charakteristischen Formen des Rechenbretts*, *Zeitschrift für Lehrmittelwesen* etc., Wien 1906, II. Jahrg., p. 47 (not historical). There is also an extensive literature relating to other forms of the abacus.

was a late product, papyrus being unknown in Greece for example before the seventh century B. C., parchment being an invention of the fifth[1] century B. C., paper being a relatively late product,[2] and metal and stone being the common media for the transmission of written knowledge in the earlier centuries in China. On account of the crude numeral systems of the ancients and the scarcity of convenient writing material, there were invented in very early times various forms of the abacus, and this instrumental arithmetic did not give way to the graphical in western Europe until well into the Renaissance period.[3] In eastern Europe it never has been replaced, for the *tschotü* is used everywhere in Russia today, and when one passes over into Persia the same type of abacus[4] is common in all the bazaars. In China the *swan-pan* is universally used for purposes of computation, and in Japan the *soroban* is as strongly entrenched as it was before the invasion of western ideas.

The Japanese *soroban* is a comparatively recent invention, having been derived from the Chinese *swan-pan* (Fig. 10), which is also relatively modern. The earlier means employed in China are known to us chiefly through the masterly work of Mei Wen-ting (1633—1721)[5] entitled *Kou-swan-k'i-k'ao.*[6] Mei Wen-ting was one of the greatest Chinese mathematicians, the author of upwards of eighty works or memoirs, and one of the leading writers on the history of mathematics among his people. He tells us that the early instrument of calculation was a set

---

1 Pliny says of the second century B. C.

2 It seems to have been brought into Europe by the Moors in the twelfth century.

3 See SMITH, D. E., *Rara Arithmetica*, Boston 1909, index under *Counters*.

4 Known in Armenia as the *choreb*, in Turkey as the *coulba*.

5 Surnamed Ting-kieou and Wou-ngan. He lived in the brilliant reign of Kang-hi, who had been educated partly under tbe influence of the Jesuit missionaries.

6 Researches on ancient calculating instruments. See VISSIÈRE, loc. cit., p. 7, from whom I have freely quoted; WYLIE, A., *Notes on Chinese Literature*, p. 91.

of rods, *ch'eou.*[1] The earliest definite information that we have of the use of these rods is in the *Han Shu* (Records of the Han Dynasty), which was written by Pan Ku of the Later Han period, in the year 80 of our era. According to him the ancient arithmeticians used comparatively long rods,[2] and the commentary of Sou Lin on the Han history tells us that two hundred seventy-one of these formed a set.[3] Furthermore, in the *Che-chouo* (Narrative of the Century), written by Lieou Yi-k'ing in the fifth century, it appears that ivory rods were used. We also find that the ancient ideograph for *swan* (reckoning) is , a form that is manifestly derived from the rods, and that is evidently the source of the present Chinese ideograph. Mei Wen-ting says that it is impossible to give the origin of these rods, but he believes that the ancient classic, the *Yih-king*, gives evidence, in its mystic trigrams, of their very early use.[4] As to the size of the rods in ancient times we are not informed, none being now extant, but an early work on cooking, the *Chong-k'ouei-lou*, speaks of cutting pieces of meat 3 inches long, like a calculating rod, from which we get some idea of their length.

As to the early Chinese method of representing numbers, we have a description by Ts'ai Ch'en, surnamed Kieou-fong (1167—1230), a philosopher of the Song dynasty. In his *Hong-fan* (Book of Annals) he gives the numerals as follows:

1 2 3 4 5 6 7 8 9 12 25 46 69 99

---

1 There is not space in this work to enter into a discussion of the possible earlier use of knotted cords, a primitive system in many parts of the world. Lao-tze, "the old philosopher", refers to them in his *Tao-teh-king*, a famous classic of the sixth century B. C., saying: "Let the people return to knotted cords (*chieng-shing*) and use them." See the English edition by Dr. P. CARUS. Chicago, 1898, pp. 137, 272, 323.

2 The text says 6 units (inches) but we do not know the length of the unit (inch) of that period.

3 The old word means, possibly, a handful.

4 The date of the *Yih-King* or Book of Changes is uncertain. It is often spoken of as *Antiquissimus Sinarum liber*, as in an edition by JULIUS MOHL, Stuttgart, 1834—9, 2 vols. It is ascribed to Fuh-hi (B. C. 3322) the fabled founder of the nation. There is an extensive literature upon the subject.

Furthermore the great astronomer and engineer of the Mongol dynasty, Kouo Sheou-kin (1281), in his *Sheou-she Li*, a treatise on the calendar, gives the number 198617 in the following form, which may be compared with the Japanese *sangi* of which we shall presently speak: 𝍠 𝍨 𝍰 𝍥 𝍩 𝍦. This plan is much older than the thirteenth century, however, for in the *Sun-tsu Suan-ching* mentioned in Chapter II, written by Sun-tsu about the third century, it is stated that the units should be vertical, the tens horizontal, the hundreds vertical, the thousands horizontal, and so on, and that for 6 one should not use six rods, since a single rod suffices for 5. These rules are repeated, almost verbatim, in the *Hia-heou Yang Suan-ching*, one of the Chinese mathematical classics, probably of the sixth century. The rods are therefore very old, and they were the common means of representing numbers in China, as we shall see was also the case in Japan, until a relatively late period.

As to the methods of operating with the rods, Yang Houei, in his *Siu-kou-Ch'ai-ki-Swan-fa* of 1275 or 1276, gives the following example in multiplication:

𝍪 𝍣 𝍯	= multiplier	=	247
𝍯 𝍢 𝍮	= multiplicand	=	736
𝍠 𝍰 𝍠 𝍯 𝍨 𝍪	= product	=	181 792

From China the calculating rods passed to Korea where the natives use them even to this day. These sticks are commonly made of bamboo, split into square prisms, and numbering about 150 in a set. They are kept in a bamboo case, although some are made of bone and are kept in a cloth bag as shown in the illustration, (Fig. 4.). The Korean represents his numbers from left to right, laying the rods as follows:

I	II	III	IIII	X	XI	XII	XIII	XIIII	—	T[1]
1	2	3	4	5	6	7	8	9	10	11

---

1 We are indebted to an educated Korean, Mr. C. Cho, of the Methodist Publishing House in Tokio, for this information. On the mathematics of Korea in general, see LOWELL, P., *The Land of the Morning Calm*. Boston 1886, p. 250. One of the leading classics of the country is the *Song yang hoei soan fa*, or *Song yang houi san pep* (Treatise on Arithmetic by Yang Hoei

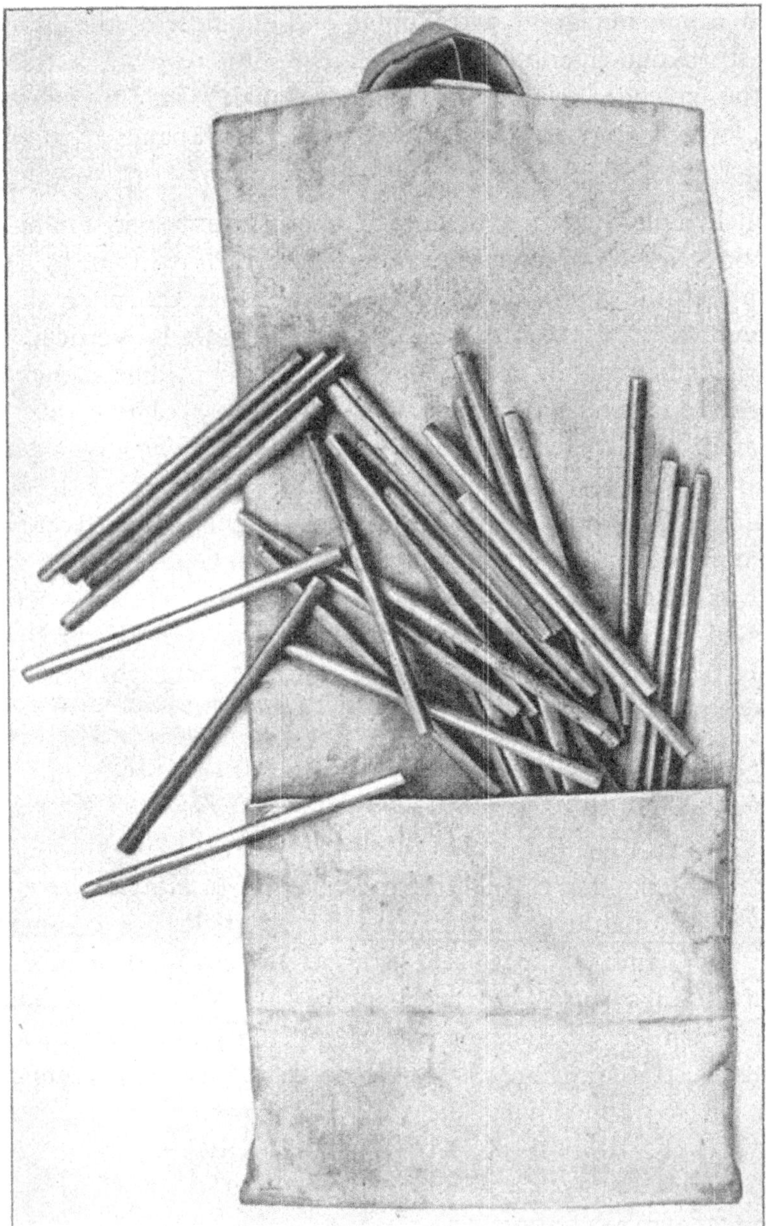

Fig. 4. Korean computing rods.

---

of the Song Dynasty), written in 1275 by Yang Hoei, whose literary name was Khien Koang; see M. COURANT, *Bibliographie Coréenne.* Paris 1896, vol. III, p. 1.

The date of the introduction of the rods into Japan is unknown, but at any rate from the time of the Empress Suiko (593—628 A. D.)[1] the *chikusaku* (bamboo rods) were used. These were thin round sticks about 2 mm. in diameter and 12 cm. in length, but because of their liability to roll they were in due time replaced by the *sangi* pieces, square prisms about 7 mm. thick and 5 cm. long. (Fig. 5.) When this transition

Fig. 5. The *sangi* or computing rods. Nineteenth century specimens.

took place is unknown, nor is it material since the methods of using the two were the same.[2]

The method of representing the numbers by means of the *sangi* was the same as the one already described as having long been used by the Chinese. The units, hundreds, ten

---

[1] HAYASHI, T, *A brief history of the Japanese Mathematics*, in the *Nieuw Archief voor Wiskunde*, tweede Reeks, zesde en sevende Deel, part I, p. 18.

[2] Indeed it is not certain that there was a sudden change from one to the other or that the names signified two different forms. The old Chinese names were *ch'eou* (which is the Japanese *sangi*) and *t'sê*, and these were used as synonymous.

thousands, and so on for the odd places, were represented as follows:

𝍠	𝍡	𝍢	𝍣	𝍤	𝍥	𝍦	𝍧	𝍨
1	2	3	4	5	6	7	8	9

The tens, thousands, hundred thousands, and so on for the the even places, were represented as follows:

𝍩	𝍪	𝍫	𝍬	𝍭	𝍮	𝍯	𝍰	𝍱
10	20	30	40	50	60	70	80	90

These numerals were arranged in a series of squares resembling our chess-board, called a *swan-pan*, although not at all like the Chinese abacus that bears this name. The following illustration (Fig. 6), taken from Satō Shigeharu's *Tengen Shinan* of 1698, shows its general form:

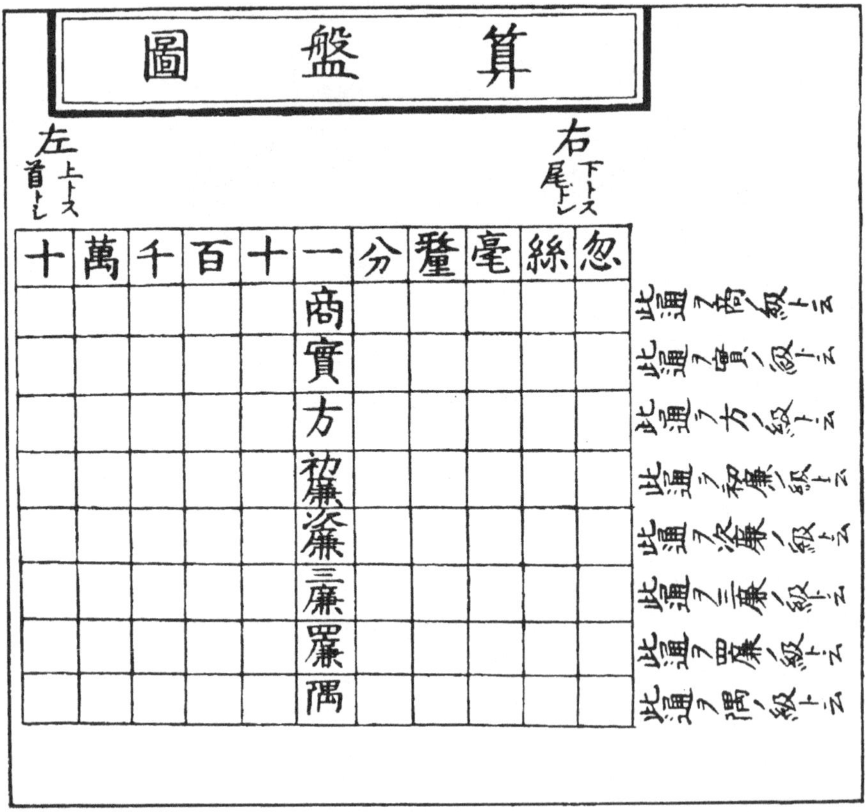

Fig. 6. The general form of the *sangi* board, from a work of 1698.

The number 38057, for example, would be represented thus:

The number 1267, represented by the *sangi* without the ruled board. Is shown in Fig. 7.

From representing the numbers by the *sangi* on a ruled board came a much later method of transferring the lines to

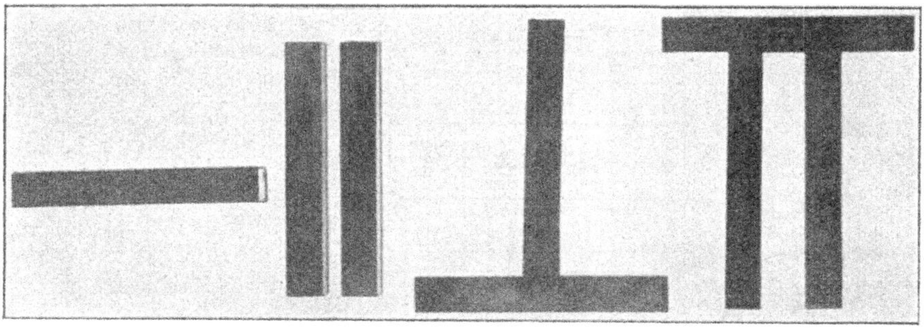

Fig. 7. The number 1267 represented by *sangi*.

paper, and using a circle to represent the vacant square. This could only have occurred after the zero had reached China and had been passed on to Japan, but the date is only a matter of conjecture. By this method, instead of having 38057 represented as shown above, we should have it written thus:

In laying down the rods a red piece indicated a positive number and a black one a negative. In writing, however, a mark placed obliquely across a number indicated subtraction. Thus,  meant — 3, and  meant — 6.

The use of the *sangi* in the fundamental operations may be illustrated by the following example in which we are required to find the *shō* (quotient) given the *jitsu* (dividend) 276, and the *hō* (divisor) 12.[1]

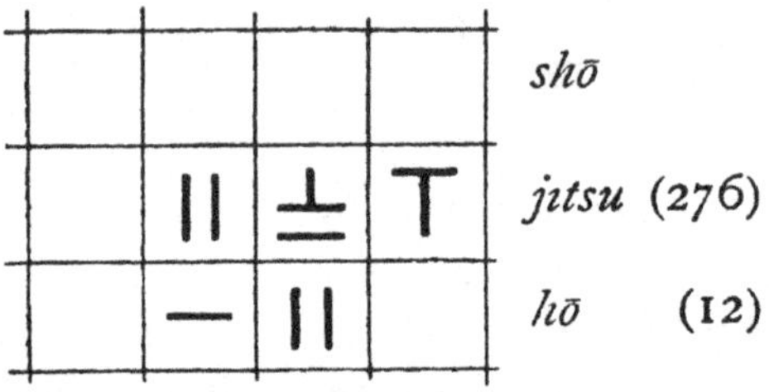

First consider the *jitsu* as negative, indicating the fact in this manner:

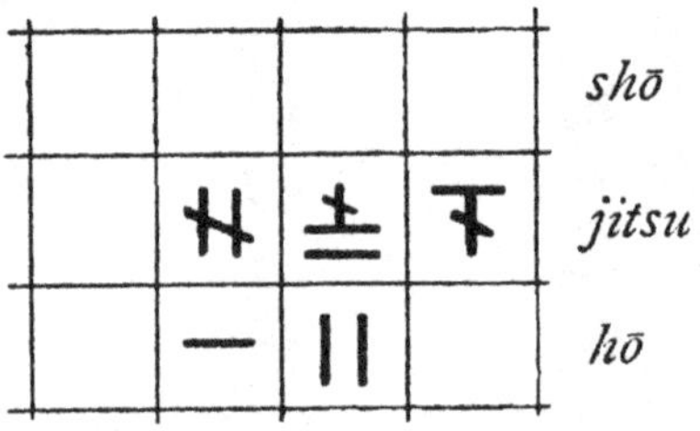

The first figure of the *shō* is evidently 2:

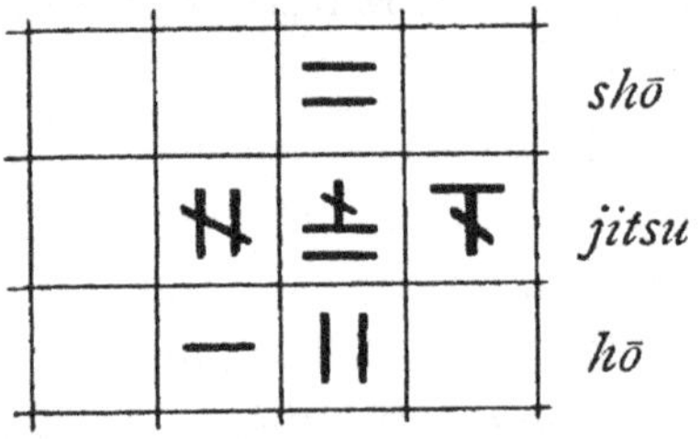

Multiply the *hō* by 20, and put the product, 240, beside the *jitsu*, thus:

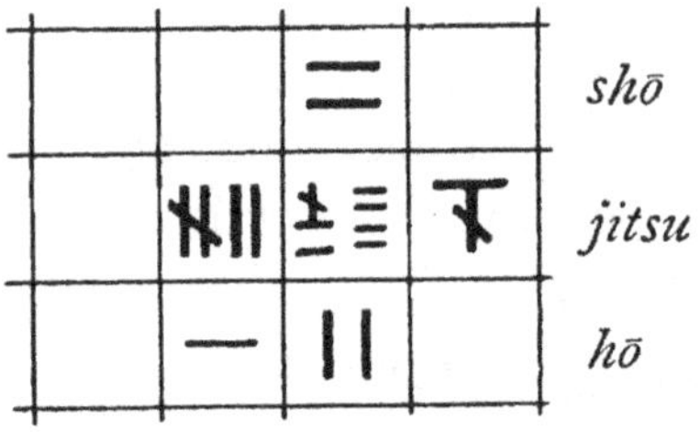

---

[1] These examples are taken from HAYASHI's *History.*

which, by combining numbers in the *jitsu*, reduces to

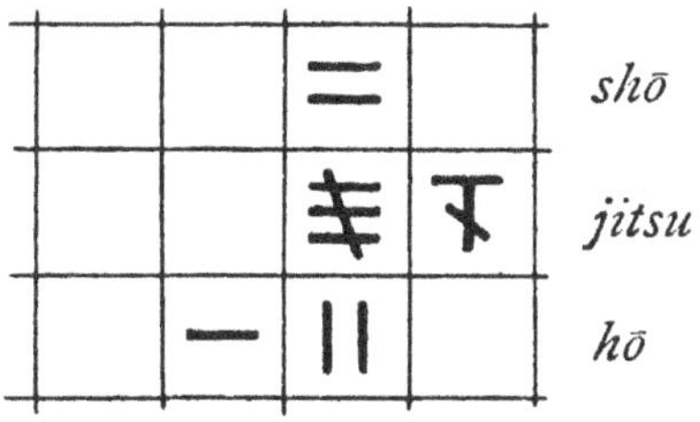

The *hō* is now advanced one place, exactly as was done in the early European plan of division by the galley method, after which the next figure of the *shō* is evidently 3, and the work appears as follows:

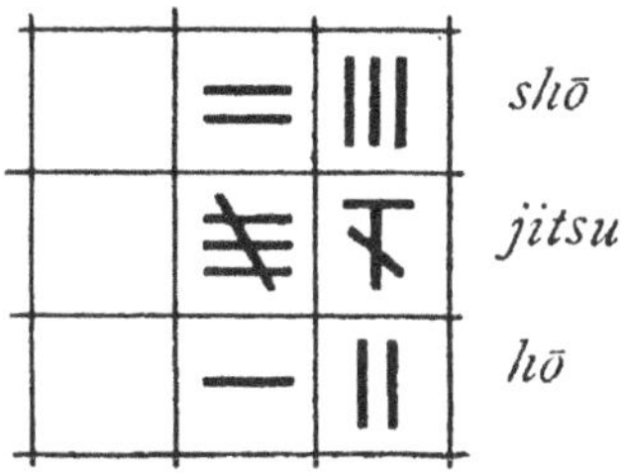

Multiplying the *hō* by 3 the product, 36, is again written beside the *jitsu*, giving

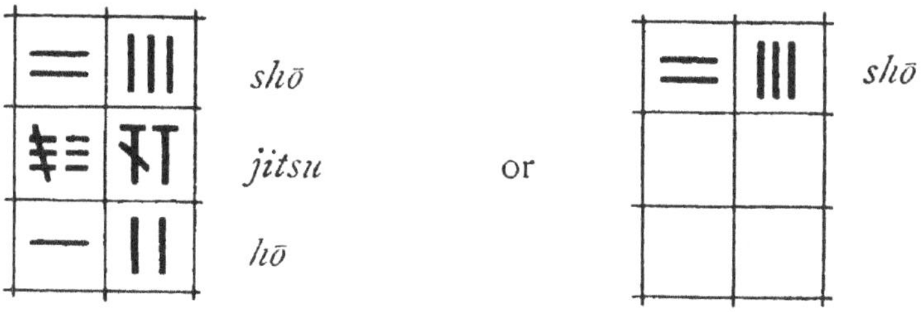

a result which is written thus: 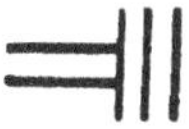 .

In order that the appearance of the *sangi* in actual use may be more clearly seen, a page from Nishiwaki Richyū's *Sampō Tengen Roku* of 1714 is reproduced in Fig. 8, and an illustration from Miyake Kenryū's *Shōjutsu Sangaku Zuye* of 1795 in Fig. 9.

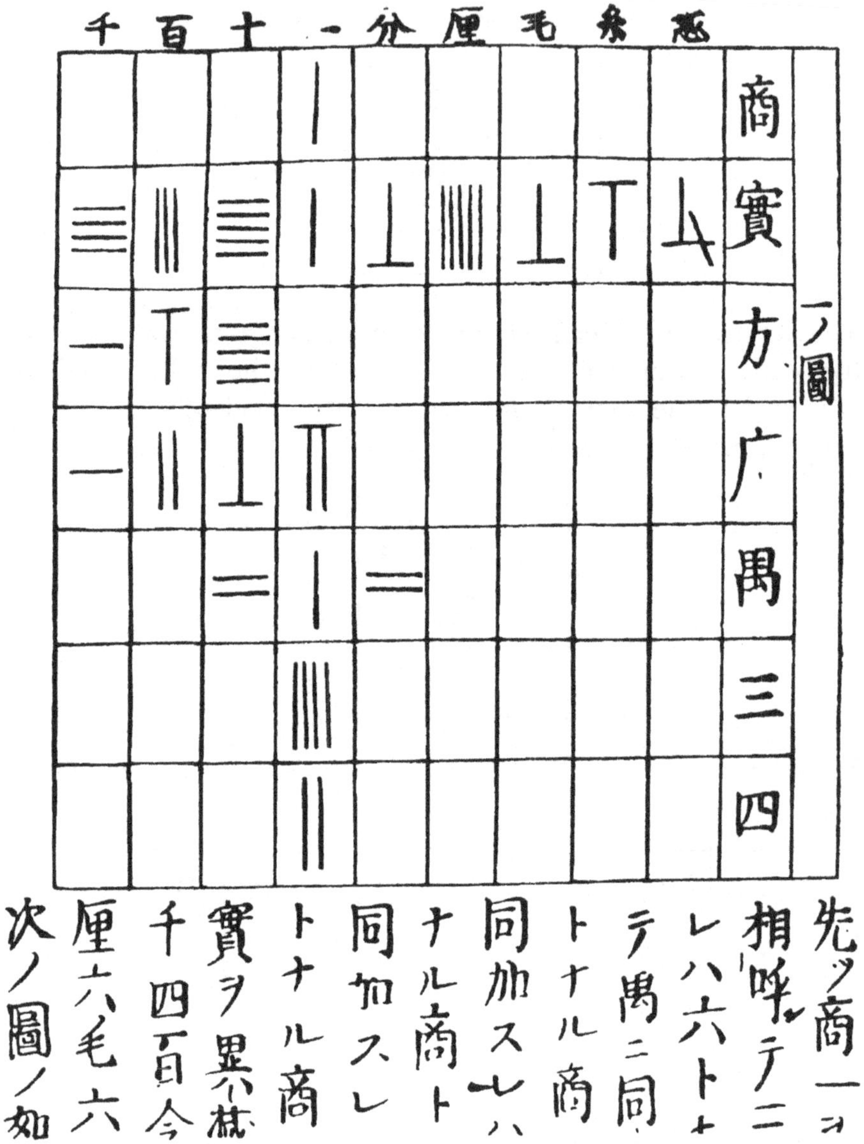

Fig. 8. Sangi board. From Nishiwaki Richyū's *Sampō Tengen Roku* of 1714.

In the later years of the *sangi* computation the custom of arranging the even places differently from the odd places changed, and instead of representing 38057 by the old method[1] as shown on page 25, it was represented thus:

1 Called *Son-shi-Reppu-hō,* the Method of arrangement of Sun-tsu.

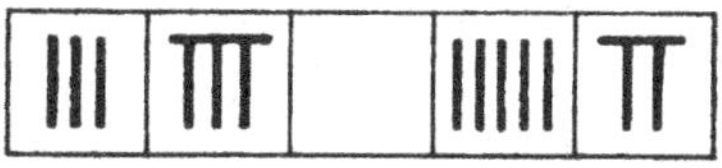

This was done only on the ruled squares, however, the written form remaining as shown on page 25.

The transition from the *ch'eou* or rod calculation to the present form of abacus in China next demands our attention. Mei Wen-ting, whose name has already been mentioned, expresses regret that an exact date for the abacus cannot be

Fig. 9. · From Miyake Kenryū's work of 1795.

fixed. He says, however, "If, in my ignorance, I may be allowed to hazard a guess, I should say that it began with the first years of the Ming Dynasty." This would be aboud 1384, when T'ai-tsou, the first Ming emperor, undertook to reform the calendar. At any rate, Mei Wen-ting concludes that in the reform of the calendar in 1281 rods were used, while in that of 1384 the abacus was employed. There is evidence, however, that the abacus was known in China in the twelfth century, but that it was not until the fourteenth that it was commonly used.[1] Since a division table such as is used in manipulating the *swan-pan* is given in a work by Yang Hui who flourished at the close of the Song Dynasty, in the latter

1 VISSIÈRE, *loc. cit.*; MIKAMI, Y., *A Remark on the Chinese Mathematics in Cantor's Geschichte der Mathematik, Archiv der Mathematik und Physik,* vol. XV (3), Heft 1.

half of the thirteenth century, we have reason to believe that the *swan-pan* was known at that time. Moreover we have the titles of several books such as *Chou-pan Chi* and *Pan-chou Chi* recorded in the Historical Records of the Song Dynasty, which seem to refer to this instrument. It must also be admitted that at least one much earlier work mentions "computations by means of balls," although this seems to have been only a

Fig. 10. The Chinese *swan-pan*, indicating the number 27091.

local plan known to but few. That the Roman abacus should have been known very early in China is not only probable but fairly certain, in view of the relations between China and Italy at the time of the Caesars.[1]

The Chinese abacus is known commonly as the *swan-pan* (*swan-p'an*, "reckoning table"). In southern China it is also known as the *soo-pan*,[2] and in Calcutta, where the Chinese shroffs employ it, the name is corrupted to *swinbon*. The literary name is *chou-p'an* ("ball table" or "pearl table"). As will be seen by the illustration there are five balls below the

---

1 See Smith and Karpinski, *loc. cit.*, p. 79.

2 Bowring, J., *The Decimal System*. London 1854, p. 193.

line and two above, each of the latter counting as five. In the illustration (Fig. 10) the balls are placed to represent 27091. The balls are called *chou* (pearls) or *tse* (son, child, grain), and are commonly spoken of as *swan-p'an chou-tse*. The transverse bar is the *leang* (beam) or *tsi-leang* (spinal colum, also used to designate the ridge-pole of a roof). The columns are called *wei* (positions), *hang* (lines), or *tang* (steps, or bars). The left side is called *ts'ien* (front) and the right side *heou* (rear). This was the instrument that replaced the ancient rods about the year 1300, perhaps suggested by the ancient Roman abacus which it resembles quite closely, perhaps by some form of instrument in Central Asia, and perhaps invented by the Chinese themselves. The resemblance to the Roman form, and the known intercourse with the West, both favor the first of these hypotheses.

Fig. 11. The *soroban*, indicating the number 90278 in the middle of the board.

Just as the Japanese received the *sangi* from China, perhaps by way of Korea, so they received the abacus from the same source. They call their nstrument by the name *soroban*, which some have thought to be a corruption of the Chinese *swan-pan*,[1] and others to have been derived from the word *soroiban*, meaning an orderly arranged table.[2]

The *soroban* is an improvement upon the *swan-pan*, as will be seen by the illustration. Instead of

---

[1] KNOTT, loc. cit., p. 45.

[2] OYAMADA, *Matsunoya Hikki*.

having two 5-balls it has only one, and it replaces the balls by buttons having a sharp edge that the finger easily engages without slipping. In the illustration (Fig. 11) the number 90278 is represented in the center of the *soroban*.

The invention of the *soroban*, or rather the importation and the improvement of the *swan-pan*, is usually assigned to the close of the sixteenth century, although we shall show that this is probably too late a date. In the *Sampō Tamatebako*, by Fukuda Riken, published in 1879, an account is given of the journey of one Mōri Kambei Shigeyoshi, a scholar of the sixteenth century, to China. Mōri was in his early days in the service of Lord Ikeda Terumasa, and was afterwards a retainer of the great hero Toyotomi Hideyoshi, better known as Taikō, who in the turbulent days of the close of the Ashikaga Shogunate[1] subdued the entire country, compelling peace by force of arms. The story goes that Taikō, wishing to make his court a center of learning, sent Mōri to China to acquire the mathematical knowledge that was wholly wanting in Japan at that period. Mōri, however, was a man of humble station, and his requests on behalf of his master were treated with such contempt that he returned to his native land with little to show for his efforts. Upon relating his trials and humiliation to Taikō, the latter bestowed upon him the title of *Dewa no Kami*, or Lord of Dewa. Again Mōri set out for China, but again he was destined to meet with some dissappointment, for hardly had he set foot on Chinese soil than Taikō began his invasion of Korea. China at once became involved in the defence of what was practically a vassal state, and as the war progressed it became more and more a matter of danger for a Japanese to reside within her borders. Mōri was not received with the favor that he had hoped for, and in due time returned to his native land. Although he had spent some time abroad, he had not accomplished his entire purpose. Nevertheless he brought back with him a considerable knowledge

---

[1] This just preceded the Tokugawa shogunate, which lasted from 1603 to 1868.

of Chinese mathematics, and also the *swan-pan*, which was forthwith developed into the present *soroban*. If the story is true, Mōri must have spent some years in China, for Taikō began his invasion in 1592 and died in 1598, and he was already dead when Mōri returned. Mōri repaired to the Castle of Ōsaka which Taikō had built and where he had lived, and there he was hospitably received by the son and successor of the great warrior. There he lived and wrote until the city was besieged in 1615, and the castle taken by Japan's greatest hero, Tokugawa Iyeyasu, founder of the Tokugawa shogunate, whose tomb at Nikkō is a Mecca for all tourists to that delightful region. We are told by Araki,[1] who lived at the beginning of the eighteenth century, that Mōri thenceforth taught the soroban arithmetic in Kyōto.

Although this story of Mōri's visit to China and of his introduction of the *soroban* is a recent one, it has been credited by some of the best writers in Japan.[2] Nevertheless there is a good deal of uncertainty about his journey,[3] and still more about his having been the one to introduce the *soroban* into Japan. Fukuda Riken who, as we have said, first published the story in 1879, gives no sources for his information. He received his information largely from his friend C. Kawakita, who tells the writers that it was Uchida Gokan who started the story of Mōri's first Chinese journey, claiming that he had read it once upon a time in a certain old manuscript that was in the library of Yushima, in Yedo. Unfortunately on the dissolution of the shogunate, at the time of the rise of

---

[1] In the *Araki Son-yei Chadan*, or Stories told by Araki (Hikoshirō) Son-yei (1640—1718).

[2] Endō, Book I, p. 45—46, 54—56; Hayashi, *History*, p. 30, and his biographical sketch of Seki Kōwa in the *Honchō Sūgaku Kōenshū* (Lectures on the Mathematics of Japan), 1908, pp. 8—10.

[3] For example, Alfred Westphal claims that it was Korea rather than China that Mōri visited. See his *Beitrag zur Geschichte der Mathematik*, in the *Mittheilungen der deutschen Gesellschaft für Natur- und Völkerkunde Ostasiens in Tōkyō*, IX. Heft, 1876. The Chinese journey is looked upon as fiction by the learned C. Kawakita, who has studied very carefully the biographies of the Japanese mathematicians.

the modern Empire, the books of this library were dispersed and the manuscript in question seems to have been irretrievably lost. That Uchida claims to have seen it we have been personally informed both by Mr. Kawakita and by Mr. N. Okamoto, to whom he told the circumstance. Nevertheless as historical evidence all this is practically worthless. Uchida was a learned man, but his reputation was not above reproach. He never told the story until the manuscript had disappeared, and no one has the slightest idea of the age, the character, or the reliability of the document. Moreover the older writers make no mention of this Chinese journey, as witness the *Araki Son-yei Chadan* which was written only a century after Mōri lived and which gives a sketch of his life and a brief statement concerning the early Japanese mathematics. In Murai's *Sampō Dōshi-mon,*[1] written nearly a century later still, no mention is made of the matter. Indeed, it is not until after the story was started by Uchida that we ever hear of it.[2]

But whether or not Mōri went to China, he did much for mathematics and he was an expert in the manipulation of the *soroban.* He was also possessed of a well-known Chinese treatise on the *swan-pan,* written by Ch'eng Tai-wei[3] and published in 1593,[4] a work that greatly influenced Japanese mathematics even long after Mōri's death. Mōri himself published a work on arithmetic in two books entitled *Kijo Ranjō*[5], and he left a manuscript on mathematics written in 1628.[6] Both have been lost, however, and of the contents of neither

---

1 Book I, chapter on the Origin of Arithmetic, published in 1781.

2 The oldest manuscript that we have found that speaks of it is SHIRAISHI's *Sūka Jimmei-Shi,* but since the author was a contemporary of Uchida he probably simply related the latter's story.

3 Erroneously given in ENDŌ as Ju Szŭ-pu. Book I, p. 45.

4 The *Suan-fa T'ung-tsong.*

5 The Kijohō method of division on the soroban, described later. See MURAI, *Sampō Dōshi-mon,* 1781, Book I; and ENDŌ, Book I, p. 45.

6 This fact is recorded in an anonymous manuscript entitled *Sanwa Zui-hitsu,* which relates that the original manuscript, signed and sealed by Mōri himself, was in the possession of a mathematician named Kubodera early in the nineteenth century.

have we any knowledge. Mōri seems to have made a livelihood after the fall of Ōsaka by teaching arithmetic in Kyōto, where hundreds of pupils flocked to learn of him and study with the man who proclaimed himself "The first instructor in division in the world." He is said to have spent his last years at Yedo, the modern Tōkyō. Three of his pupils,[1] Yoshida Kōyū, Imamura Chishō, and Takahara Kisshu, known to their contemporaries as "The three Arithmeticians,"[2] did much to revive the study of the science in what we have designated as the third period of Japanese mathematics, and of them we shall speak more at length in a later chapter.

There are various reasons for believing that the *swan-pan* was not first brought to Japan by Mōri. In the first place, such simple devices of the merchant class usually find their way through the needs of trade rather than through the efforts of the scholar. It was so with the Hindu-Arabic numerals in the West,[3] and it was probably so with the *swan-pan* in the East. There is a tradition that another Mori,[4] Mori Misaburō, an inhabitant of Yamada in the province of Ise, owned a *swan-pan* in the Bun-an Era, *i. e.*, in 1444–1449. This instrument is still preserved and is now in the possession of the Kitabatake family.[5] It is also related that the great general and statesman Hosokawa Yūsai, in the time of Taikō, owned a small ivory *soroban*, but of course this may have come from his contemporary Mōri Kambei. It is, however, reasonable to believe that, with the prosperous intercourse between China and Japan during the Ashikaga Shogunate, from the fourteenth to the end of the sixteenth centuries the *swan-pan* could not have failed to become known to the Japanese merchants, even if it was not extensively used by them. On the other hand, Mōri Kambei was the first great teacher of the art of manipulating it,

---

1 See ENDŌ, Book I, p. 55, and the *Araki Son-yei Chadan.*

2 Also as the *San-shi,* or "three honorable scholars."

3 See SMITH and KARPINSKI, *loc. cit.*, p. 114.

4 Not Mōri, however.

5 It was exhibited not long ago in Tōkyō. We are indebted for this information to Mr. N. OKAMOTO.

so that much credit is due to him for its general adoption. We may, therefore, fix upon about the year 1600 as the beginning of the use of the soroban, and the century from 1600 to 1700 as the period in which it replaced the ancient bamboo rods.

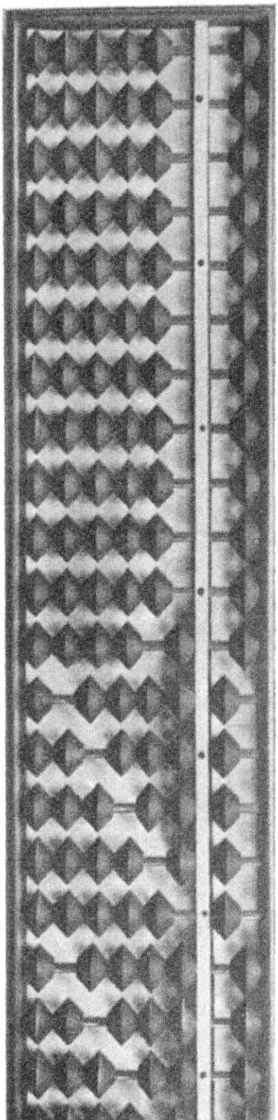

Fig. 12. The *soroban*, indicating the number 987654321.

It is proper in this connection to give a brief description of the *soroban* and of the method of operating with it, particularly with a view to the needs of the Western reader. As already stated, the value of the ball above the beam is five, one being the value of each ball below the beam. In Fig. 12 the right-hand column has been used to represent units, the next one tens, and so on. In the picture these columns have been numbered by arranging the balls so that the units are 1, the tens 2, the hundreds 3, and so on. As a result, the number represented is 987654321.[1]

To add two numbers we have only to set down the first as in the illustration and then set down the second upon it. Thus to add 2 and 2, we put 2 balls at the top of the colunn and then 2 more, making 4. To add 2 and 3, we put 2 balls at the top, and then add 3; but since this makes 5 we push back the 5 balls and move down the one above the beam. To add 4 and 3, we take 4 balls; then we add the 3 by first adding 1, moving down the one above the beam to replace the 5, and then

[1] The best description of this instrument, in English, is that given by Knott, *loc. cit.*, p. 45.

adding 2 more, leaving the five-ball and 2 unit balls. To add 7 and 6, we set down the 7 by moving the five-ball and 2 unit balls; we then move 3 more balls, which give us 10, and we indicate this by moving 1 ball in tens' column, clearing the units' column at the same time, and then we add 3 more, making 1 ten and 3 units. It will be seen that as fast as any number is set down it is thereby added to the preceding sum, thus making the work very rapid in the hands of a skilled operator. Subtraction is evidently performed with equal ease.

For multiplying readily on the *soroban* it is necessary to learn the multiplication table. In this table the Japanese have two points of advantage over the Western peoples: (1) they do not use the words "times" or "equals", thus saving considerably in time and energy whenever they employ it; (2) they learn their products only one way, as 6 7's but not 7 6's. Thus their table for 6 is as follows:[1]

Japanese names			*In our figures*		
ichi	roku	roku[2]	1	6	6
ni	roku	jū ni	2	6	12
san	roku[3]	jū hachi	3	6	18
shi	roku	ni jū shi	4	6	24
go	roku	san jū	5	6	30
roku	roku	san jū roku	6	6	36
roku	shichi	shi jū ni	6	7	42
roku	hachi[4]	shi jū hachi	6	8	48
roku	ku[5]	go jū shi	6	9	54

This table reminds us of the one in common use by the Italian merchants from the fourteenth to the sixteenth century, and which was probably quite universal in the mercantile houses.

For purposes of historic interest we take to illustrate the process of multiplication an example from the *Jinkō-ki* of

[1] KNOTT, *loc. cit.*, p. 50.

[2] This is usually stated as "*in roku ga roku*," the *ichi* being corrupted to *in* and the *ga* inserted for euphony.

[3] Corrupted to *sabu roku*.

[4] The *hachi* is abbreviated to *ha* in this case, for euphony.

[5] *Roku ku* may here be abbreviated to *rokku*.

Yoshida, published in 1627, and described more fully in Chapter V. To multiply 625 by 16 the multiplier is placed to the left of the multiplicand on the *soroban*, a plan that is exactly opposite to the Chinese arrangement as set forth in the *Suan-fa T'ung-tsong* of 1593. It represents one of the im-

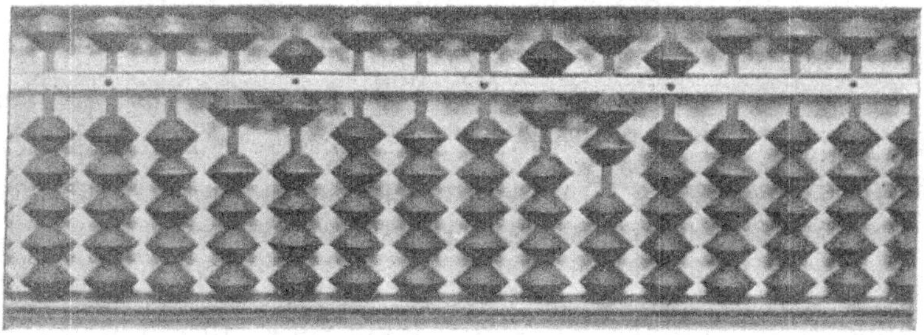

Fig. 13. 16 625.

provements of Mōri or of Yoshida, and has always been followed in Japan.

We first take the partial product $5 \times 6 = 30$, and place the 30 just to the right of the 625,[1] so that the *soroban* reads

16 62530

Fig. 14. 16 62530.

We now take $5 \times 1 = 5$, and add this 5 to the 3, making the product 80 thus far. The 5 of the 625 now having been

[1] In general, the units' figure of this product is placed as many columns to the right as there are figures in the multiplier.

multiplied by 16, it is removed, so that the figures stand as follows:

16 62080

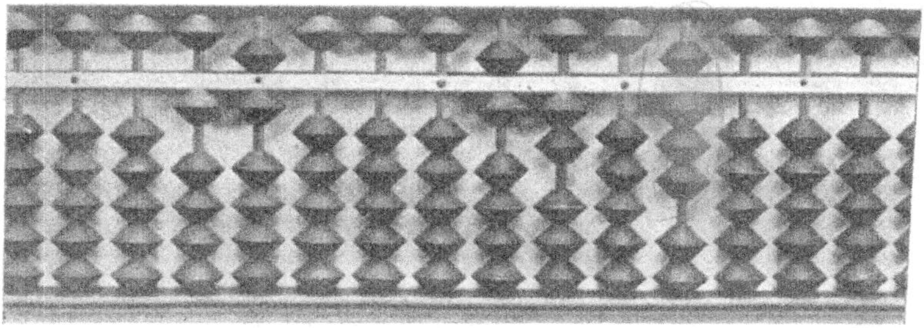

Fig. 15. 16 62080.

The next step is the multiplication of 2 by 16, and this is done precisely as the 5 was multiplied. Expressed in figures the operation on the *soroban* is as follows:

	16 62080
$2 \times 6 =$	12
$2 \times 1 =$	2
Cancel 2	16 60400

the 2 in 62080 being removed because the multiplication of 2 by 16 has been effected.

Fig. 16. 16 60400.

The next step is the multiplication of 6 by 16, and the work appears on the *soroban* as follows:

	16 60400
$6 \times 6 =$	36
$1 \times 6 =$	6
	16 10000

The result is therefore 10000.

Fig. 17. 16 10000.

The process of division is much more complicated, and requires the perfect memorizing of a table technically known as the *Ku ki hō*, or "Nine Returning Method." It is given here only for 2, 6, and 7.[1]

Ni ichi ten saku no go	2	1	replace by 5
Nitchin in jū[2]	2	2	gives 1 ten
Ni shi shin ga ni jū	2	4	gives 2 tens
Ni roku shin ga san jū	2	6	gives 3 tens
Ni hachi shin ga shi jū	2	8	gives 4 tens

Table for 6.

Roku ichi kakka no shi	6	1	14
Roku ni san jū no ni	6	2	32
Roku san ten saku no go	6	3	50
Roku shi roku jū no shi	6	4	64
Roku go hachi jū no ni	6	5	82
Roku chin in jū	6	6	gives 1 ten

Table for 7.

Shichi ichi kakka no san	7	1	13
Shichi ni kakka no roku	7	2	26
Shichi san shi jū no ni	7	3	42
Shichi shi go jū no go	7	4	55
Shichi go shichi jū no ichi	7	5	71
Shichi roku hachi jū no shi	7	6	84
Shichi chin in jū	7	7	gives 1 ten

1 KNOTT, *loc. cit.*, as corrected by Mr. MIKAMI.

2 This and some others are given in the usual abridged form.

The table is not so unintelligible as it seems to a stranger, and in fact its use has certain advantages over Western methods. In the first place it is not encumbered with such words as "divided by" or "contained in," and in the second place it is not carried beyond the point where the dividend number as expressed in the table equals the divisor. It is in fact merely a table of quotients and remainders. Consider, for example, the table for 7. This states that

10 : 7 = 1, and 3 remainder
20 : 7 = 2, and 6 remainder
30 : 7 = 4, and 2 remainder
40 : 7 = 5, and 5 remainder
50 : 7 = 7, and 1 remainder
60 : 7 = 8, and 4 remainder
70 : 7 = 10

Taking again an example from the classical work of Yoshida, let us divide 1234 by 8. These numbers will be represented on the *soroban* in the usual way, and placed as follows:

8 1234

The table now gives "8 1 12", meaning that 10 : 8 = 1, with a remainder 2. We therefore leave the 1 untouched and add 2 to the next figure, the numbers then appearing as follows:

8 1434

where the 1 represents the first figure in the quotient, and 434 represents the next dividend.

The table now tells us "8 4 50", meaning that 40 : 8 = 5, with no remainder. We therefore remove the first 4 and put 5 in its place, the *soroban* now indicating

8 1534

where 15 represents the first two figures in the quotient, and 34 represents the next dividend.

The table now tells us "8 3 36", meaning that 30 : 8 = 3, with a remainder 6. This means that the next figure of the quotient is 3, and that we have 6 + 4 still to divide. The *soroban* is therefore arranged to indicate

8 153 (10)

But 10 : 8 = 1, with a remainder 2, so the *soroban* is arranged to indicate 8 1542

meaning that the quotient is 154 and the remainder is 2. We may now consider the result is 154 1/4, or we may continue the process and obtain a decimal fraction.

If the divisor has two or more figures it is convenient to have the following table in addition to the one already given:

1	with	1,	make	it	91
2	„	2,	„	„	92
3	„	3,	„	„	93
4	„	4,	„	„	94
5	„	5,	„	„	95
6	„	6,	„	„	96
7	„	7,	„	„	97
8	„	8,	„	„	98
9	„	9,	„	„	99

This means that 10 : 1 = 9 and 1 remainder, 20 : 2 = 9 and 2 remainder, and so on.

We shall sketch briefly the process of dividing 289899 by 486 as given by Yoshida. Arrange the *soroban* to indicate

486 289899.

The table gives "4 2 50", so we change the 2 to 5 and arrange the *soroban* to indicate the following:

	486	589899
$5 \times 8 =$		40
$5 \times 6 =$		30
	486	546899

Here 5 is the first figure of the quotient and 46899 is the remainder to be divided. Looking now at the last table we find "4 4 94", so we change the 4 to 9 and add 4 to the following digit. The *soroban* is arranged to indicate the following:

	486	546899
Then	486	596899
Add 4		4
Then $9 \times 8 =$		72
$9 \times 6 =$		54
Subtract 72 and 54	486	593159

Here 59 is the first part of the quotient and 3159 is the remainder to be divided.

Proceeding in the same way, the next figure in the quotient is 6, and the *soroban* indicates

486 596759
486 596243
486 5965

and the quotient is 596.5.

Fig. 18. From the work of Fujiwara Norikaze, 1825.

This method of division is that given in the *Jinkō-ki*, but in 1645 another plan was suggested by a well-known teacher, Momokawa Chubei.[1] This was the *Shōjoho*, or method of division by the aid of the ordinary multiplication table, as in written arithmetic. Momokawa sets it forth in a work entitled

1 Endō gives his personal name as Jihei, but this is open to doubt.

Kamei-zan (1645), and thenceforth the method itself bore this name. This plan, like the Jinkōki, is fundamentally a Chinese

Fig. 19. From an anonymous *Kwaisanki* of the seventeenth century.

method, as it appears in the *Suan fa T'ung-tsong* of 1593, but it has never been so popular in Japan as the one given by Yoshida in the Jinkōki.

It is hardly worth while to consider the method of extracting roots by the help of the *soroban*, since the general theory does not differ from the one used in the West, and the subsidiary operations have been sufficiently explained.

Although the *soroban* began to replace the bamboo rods soon after 1600, it took more than a century for the latter to disappear as means for computation, and, as we shall see, they continued to be used for about two hundred years longer in connection with algebraic work. In Isomura Kittoku's *Sampō Ketsugi-shō* of 1660 (second edition 1684), and Sawaguchi's *Kokon Sampō-ki* of 1670, for example, we find both the rods

Fig. 20. From Miyake Kenryū's work of 1795.

and the *soroban* explained, and in another work of 1693 only the rods are given. The *Tengen Shinan*, by Satō Shigeharu, printed in 1698, also gives only the rods, as does the *Kwatsuyō Sampō* (Method of Mathematics) which Araki Hikoshirō Sonyei, being old, caused his pupil Ōtaka Yoshimasa to prepare in 1709.[1] In Murata Tsushin's *Wakan Sampō*, published in 1743, both systems are used, and in a primary arithmetic printed in 1781 only the rods are employed, so that we see that it was a long time before the *soroban* completely replaced the more ancient method of computation. In general we may say that all algebras used the *sangi* in connection with the "celestial element" method of solving equations, explained in the next chapter, while little by little the *soroban* replaced them

[1] It was printed in 1712.

for arithmetical work. The pictures of children learning to use the *soroban* are often interesting, as in the one from the arithmetic of Fujiwara Norikaze, of 1825 (Fig. 18). The early pictures of the use of the instrument in mercantile affairs are also curious, as in Fig. 19, taken from an anonymous work of the seventeenth century. An illustration of a pupil learning the use of the *soroban*, from Miyake Kenryū's work of 1795[1] is shown in Fig. 20.

---

[1] The first edition was 1716.

CHAPTER IV.

The Sangi applied to Algebra.

As stated in the preceding chapter, it seems necessary to break the continuity of the historical narative by speaking of the introduction of the *soroban* and the *sangi*, since these mechanical devices must be known, at least in a general way, before the contributions of the later writers can be understood. As already explained, the *chikusaku* or "bamboo rods" had been brought over from China at any rate as early as 600 A. D., and for a thousand years had held sway in the domain of calculation. They had formed one of the inheritances of the people, and the fact that they are still used in Korea shows how strong their hold would naturally have been with a patriotic race like the Japanese. We have much the same experience in the Western World in connection with the metric system today. No one doubts for a moment that this system will in due time be commonly used in England and America, the race for world commerce deciding the issue even if the merits of the system should fail to do so. Nevertheless such a change comes only by degrees in democratic lands, and while our complicated system of compound numbers is rapidly giving way, the metric system is not so rapidly replacing it.

So it was in Japan in the 17th century. The *samurai* despised the plebeian *soroban*, and the guild of learning sympathized with this attitude of mind. The result was that while the *soroban* replaced the rods for business purposes, the latter maintained their supremacy in the calculations of higher mathematics.

There was a further reason for this attitude of mind in the fact that the rods were already in use in the solution of the equation, having been well known for this purpose ever since Ch'in Chiu-shao (1247), Li Yeh (1248 and 1257), and Chu Chi-chieh (1299)[1] had described them in their works.

As stated in Chapter III, the early bamboo rods tended to roll off the table or out of the group in which they were placed. On this account the Koreans use a trianguloid prism as shown in the illustration on page 22, and the Japanese in due time resorted to square prisms about 7 mm. thick and 5 cm. long. These pieces had the name *sanchu*, or, more commonly, *sangi*, and part of each set was colored red and part black, the former representing positive mumbers and the latter negative. A set of these pieces, now a rarity even in Japan, is shown on page 23.

This distinction between positive and negative is very old. In Chinese, *chêng* was the positive and *fu* the negative, and the same ideographs are employed in Japan today, only one of the terms having changed, *sei* being used for *chêng*. These Chinese terms are found in the *Chiu-chang Suan-shu* as revised by Chang T'sang in the 2nd century B. C.,[2] and hence are probably much more ancient even than the latter date. The use of the red and black for positive and negative is found in Liu Hui's commentary on the *Chiu-chang*, written in 263 A. D.,[3] but there is no reason for believing that it originated with him. It is probably one of the early mathematical inheritances of the Chinese the origin of which will never be known. As applied to the solution of the equation, however, we have no description of their use before the work of Ch'in Chiu-shao in 1247. In the treatises of Li Yeh and Chu Chi-chieh[4] there is given a method known as the *t'ien-yüen-shu*, or *tengen jutsu*

---

[1] Chu Shi-chieh, or Choo Shi-ki. Takebe's commentary (1690) upon his work of 1299 is mentioned in Chapter VII. He also wrote in 1303 a work entitled *Sze-yuen yuh-kien*, "Precious mirror of the four elements," but this is not known to have reached Japan.

[2] See No. 8 of the list described in Chap. II, p. 11.

[3] See p. 11.

[4] His work was known as *Suan-hsiao Chi-mêng*, or *Swan-hsüch-chi-mong*. It was lost to the Chinese for a long time, but Lo Shih-lin discovered a Korean edition of 1660 and reprinted it in 1839.

as it has come into the Japanese, a term meaning "The method of the celestial element."

These three writers appeared in widely separated parts of China, under the contending monarchies of Song and Yüan, at practically the same time, in the 13th century.[1] The first, Ch'in Chiu-shao,[2] introduced the Monad as the symbol for the unknown quantity, and solved certain equations of the 6th, 7th, 8th, and even higher degrees. The ancient favorite of the West, the problem of the couriers, is among his exercises. He states that he was from a province at that time held by the Yüan people (the Mongols).

The second of this trio, Li Yeh,[3] wrote "The mirror of the mensuration of circles" in which algebra is applied to trigonometry.[4] The third of the group is Chu Chi-chieh, to whose work we have just referred. That other writers of prominence had treated of algebra before this time is evident from a passage in the preface of Chu Chi-chieh's work. In this he refers to Chiang Chou Li Wend, Shih Hsing-Dao, and Liu Ju-Hsieh as having written on equations with one unknown quantity; to Li Te Tsi, who used equations with two unknowns, and to Liu Ta Chien, who used three unknowns. Chu Chi-chieh[5] seems to have been the first Chinese writer to treat of systems of linear equations with four unknowns, after the old "Nine Sections."

---

[1] WYLIE, A., *Chinese Researches*, Shanghai, 1897, Part III, p. 175; MIKAMI, Y., *A Remark on the Chinese Mathematics in Cantor's Geschichte der Mathematik*, *Archiv der Math. und Physik*, vol. XV (3), Heft 1.

[2] Tsin Kiū-tschau, Tsin Kew Chaou. His work, entitled *Su-shu Chiu-chang*, or *Shu hsüeh Chiu Chang*, appeared in 1247. He also wrote the *Shu shu ta Lueh* (General rules on arithmetic).

[3] Or Li-yay, Li was the family name, and Yeh or Yay the personal name, this being the common order. He is also known by his familiar name, Jin-king, and also as Li Ching Chai.

[4] His two works are entitled *T'sê-yüan Hai-ching* (1248) and *I-ku Yen-tuan* (1257). The dates are a little uncertain, since Li Yeh states in the preface that the second work was printed 11 years after the first. *Tse-yüan* means "to measure the circle", and *Hai-ching* means "mirror of sea".

[5] For a translation of his work I am indebted to Professor Chen of Peking University. D. E. S.

In order that we may have a better understanding of the basis upon which Japanese algebra was built, a few words are necessary upon the state to which the Chinese had brought the science by this period. While algebra had been known before the 13th century, it took a great step forward through the labors of the three men whose names have been mentioned. They called their method by various names, but the one already given, and *Lih-tien-yüen-yih*, "The setting up of the Celestial Monad", are the ones commonly used.

In general in this new algebra, unity represents the unknown quantity, and the successive powers are indicated by the place, the *sangi* being used for the coefficients, thus:

$$x^3$$
$$+ 15x^2$$
$$+ 66x$$
$$- 360$$

Li Yeh puts the absolute term on the bottom line as here shown, in his work of 1248. In his work of 1259 and in the works of Ch'in and Chu it is placed at the top. The symbol after 66 was called *yüen* and indicated the monad, while the one after 360 was called *tai*, a shortened form of *tai-kieh*, "the extreme limit". In practice they were commonly omitted. The circle is the zero in 360, and the cancellation mark indicates that the number is negative, a feature introduced by Li Yeh. With the *sangi*, red rods would be used for 1, 15, and 66, and black ones for 360. It will be noticed that this symbolism is in advance of anything that was being used in Europe at this time, and that it has some slight resemblance to that used by Bhaskara, in India, in the 12th century.

Ch'in Chiu-shao (1247) gives a method of approximating the roots of numerical higher equations which he speaks of as the *Ling-lung-kae-fang*, "Harmoniously alternating evolution", a plan in which, by the manipulation of the *sangi*, he finds the root by what is substantially the method rediscovered by Horner, in England, in 1819. Another writer of the same period, Yang Hwuy, in his analysis of the *Chiu-chang*,[1] gives the same rule under the name of *Tsang-ching-fạng*, "Accumulating involution", but he does not illustrate it by solved problems. We are therefore compelled to admit that Horner's method is a Chinese product of the 13th century, and we shall see that the Japanese adopted it in what we have called the third period of their mathematical history.

It is also interesting to know that Çhu Chi-chieh in the *Sze-yüen Yu-kien* (1303) gives as an "ancient method" the relation of the binomial coefficients known in Europe as the "Pascal triangle",[2] and that among his names for the various monads (unknowns) is the equivalent for *thing*.[3] This is the same as the Latin *res* and the Italian *cosa*, both of which had undoubtedly come from the East. It is one of the many interesting problems in the history of mathematics to trace the origin of this term.[4]

Chu Chi-chieh writes the equivalent of $a + b + c + x$ as is here shown, except that we use T for the symbol *tai*, and the modern numerals instead of the *sangi* forms.

	1	
1	T	1
	1	

The square of this expression he writes thus:

		1		
	2	0	2	
1	0	T ^{2}	0	1
	2 ^{2}	0	2	
		1		

a method that is quickly learned and easily employed.

---

1 See p. 11.

2 This was also known in Europe long before Pascal. See SMITH, D. E., *Rara Arithmetica*, Boston, 1909, p. 156.

3 He uses the names *heaven*, *earth*, *man*, *thing*, although the first three usually designated known quantities.

4 The resemblance to the Egyptian *ahe*, mass (or *hau*, heap), of the Ahmes papyrus, c. 1700 B. C., will possibly occur to the reader.

The "celestial element" process as given by Chu Chi-chieh in 1299 found its way into Japan at least as early as the middle of the 17th century, and the *Suan-hsiao Chi-mêng* was reprinted there no less than three times.[1] The single rule laid down in this classical work for the use of the *sangi* in the solution of numerical equations contains but little positive information. Retaining the Japanese terms, and translating quite literally, we may state it as follows:—

"Arrange the *seki* in the *jitsu* class, adjusting the *hō*, *ren*, and *gū* classes. Then add the like-signed and subtract the unlike-signed, and evolve the root."

This reminds one of the cryptic rules of the Middle Ages and early Renaissance in Europe, but unlike some of these it is at least not an anagram to which there is no key. The *seki* is the quantity in a problem that must be expressed in the absolute term before solving, and which is represented by the *sangi* in next to the top row, the *jitsu* class. The coefficients of the first, second, and third powers of the unknown are then represented by the *sangi* in the successive rows below, in the *hō*, *ren*, and *gū* classes. The rest of the rule amounts to saying that the pupil should proceed as he has been taught. The method is best understood by actually solving a numerical higher equation, but inasmuch as the manipulation of the *sangi* has already been explained in the preceding chapter, the coefficients will now be represented by modern numerals. The problem which we shall use is taken from the eighth book of the *Tengen Shinan* of Satō Moshun or Shigeharu, published in 1698, and only the general directions will be given, as was the custom. The reader may compare the work with the common Horner method in which the reasoning involved is more clear.

Let it be required to solve the equation

$$11\,520 - 432x - 236x^2 + 4x^3 + x^4 = 0$$

[1] For the first time in 1658. Dōwun, a Buddhist priest, with the possible *nom de plume* of Baisho, mentions one Hisada (or Kuda) Gentetsu (probably also a priest) as the editor. It was also printed in 1672 by Hoshino Jitsusen, and some time later by Takebe Kenkō.

Arrange the sangi on the board to indicate the following:

(r)					
(0)	1	1	5	2	0
(1)		—	4	3	2
(2)		—	2	3	6
(3)					4
(4)					1

Here the top line, marked (r), is reserved for the root, and the lines marked (0), (1), (2), (3), (4) are filled with the *sangi* representing the coefficients of the 0th, 1st, 2d, 3d, and 4th powers of the unknown quantity. With the *sangi*, the negative 432 and 236 would be in black, while the positive 11520, 4, and 1 would be in red.

First advance the 1st, 2d, 3d, and 4th degree classes 1, 2, 3, 4 places respectively, thus:

(r)					
(0)	1	1	5	2	0
(1)	—	4	3	2	
(2)	– 2	3	6		
(3)		4			
(4)	1				

The root will have two figures and the tens' figure is 1. Multiply this 10 by the 1 of class (4) and add it to class (3), thus making 14 in class (3). Multiply this 14 by the root, 10, and add it to —236 of class (2), thus making —96 in class (2). Multiply this —96 by the root, 10, and add it to —432 of class

(1), thus making —1392 in class (1). Multiply this —1392 by the root, 10, and add it to 11 520 of class (0), thus making —2400. The result then appears as follows:

(r)				1	
(0)		−2	4	0	0
(1)	−1	3	9	2	
(2)		−9	6		
(3)	1	4			
(4)	1				

Now repeat the process, multiplying the root, 10, into class (4) and adding to class (3), making 24; multiply 24 by the root and add to class (2), making 144; multiply 144 by the root and add to class (1), making 48. The result then appears as follows:

(r)				1	
(0)		−2	4	0	0
(1)			4	8	
(2)	1	4	4		
(3)	2	4			
(4)	1				

Repeat the process, multiplying the root, 10, into class (4) and adding to class (3), making 34; multiply 34 by the root and add to class (2) making 484.

Again repeat the process, multiplying the root into class (4) and adding to class (3), making 44.

Now move the *sangi* representing the coefficients of classes

(1), (2), (3), (4), to the right 1, 2, 3, 4, places, respectively, and we have:

(r)				1	
(0)		−2	4	0	0
(1)				4	8
(2)			4	8	4
(3)				4	4
(4)					1

The second figure of the root is 2.[1] Multiply this into class (4) and add to class (3), making 46. Multiply the same root figure, 2, into this class (3) and add to class (2), making 576. Multiply this 576 by 2 and add to class (1), making 1200. Multiply this 1200 by 2 and add to class (0), making 0. The work now appears as follows:—

(r)				1	2
(0)					
(1)		1	2		
(2)			5	7	6
(3)				4	6
(4)					1

The root therefore is 12.

It may now be helpful to give a synoptic arrangement of the entire process in order that this Chinese method of the 13th century, practiced in Japan in the 17th century, may be

[1] It is not stated how either figure is ascertained.

compared with Horner's method. The work as described is substantially as follows:

Given $x^4 + 4x^3 - 236x^2 - 432x + 11\,520 = 0$

1	+ 4	— 236	— 432	+ 11520
	10	140	— 960	— 13920
1	14	— 96	— 1392	— 2400
	10	240	1440	
1	24	144	48	— 2400
	10	340		2400
1	34	484	48	0
	10		1152	
1	44	484	1200	
	2	92		
1	46	576		

Chu Chi-chieh also gives, in the *Suan-hsiao Chi-mêng*, rules for the treatment of negative numbers. The following translations are as literal as the circumstances allow:

"When the same-named diminish each other, the different-named should be added together.[1] If then there is no opponent for a positive term, make it negative; and for a negative, make it positive."[2]

"When the different-named diminish each other the same-named should be added together. If then there is no opponent for a positive, make it positive; and for a negative, make it negative."[3]

"When the same-named are multiplied together, the product is made positive. When the different-named are multiplied together, the product is made negative."

The method of the "celestial element", with the *sangi,* and with the rules just stated, entered into the Japanese mathe-

---

[1] This is intended to mean that when $(+4) - (+3) = +(4-3)$, then $(+4) - (-3)$ should be $+4+3$.

[2] That is, $0 - (+4) = -4$, and $0 - (-4) = +4$.

[3] When $(+p) - (-q) = +p + q$, then $(-p) - (+q) = -(p+q)$. Also, $0 + (+4) = +4$, and $0 + (-4) = -4$.

matics of the 17th century, to be described in the following chapter. They were purely Chinese in origin, but Japan advanced the method, carrying it to a high degree of perfection at the time when China was abandoning her native mathematics under the influence of the Jesuits. It is, therefore, in Japan rather than China that we must look in the 17th century for the strictly oriental development of calculation, of algebra, and of geometry.

Among the other writers of the period several treated of magic squares. Among these was Hoshino Sanenobu, whose *Kō-ko-gen Shō* (Triangular Extract) appeared in 1673. Half of one of his magic squares in shown in the following facsimile:

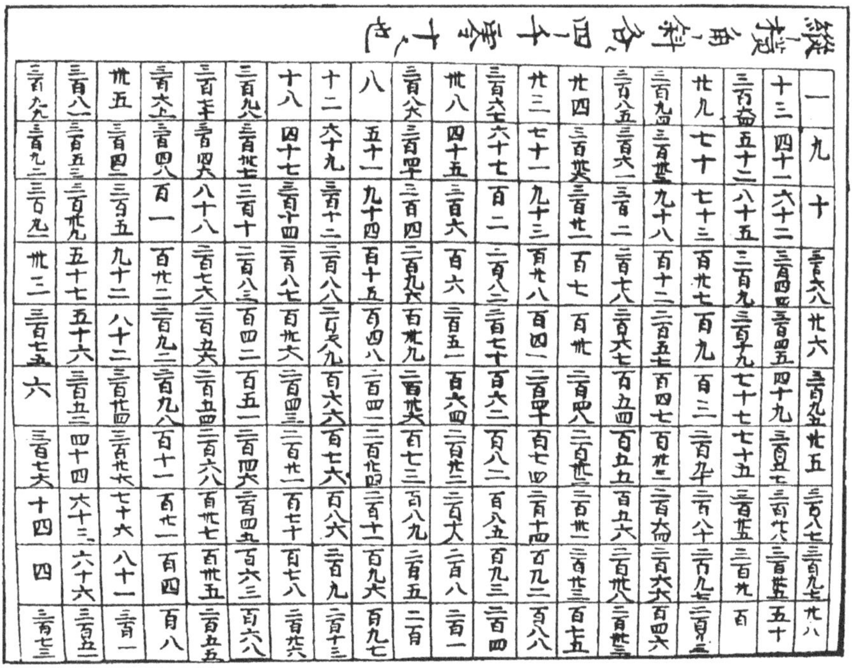

Fig. 21. Half of a magic square, from Hoshino Sanenobu's work of 1673.

One who is not of the Japanese race cannot refrain from marvelling at the ingenuity of many of these problems proposed during the 17th century, and at the painstaking efforts put forth in their solution. He is reminded of the intricate ivory carvings of these ingenious and patient people, of the curious puzzles with which they delight the world, and of the finish which characterizes their artistic productions. Few of these problems could be mistaken for western productions, and the solutions, so far as they are given, are like the art and the literature of the people, indigenous to the soil of Japan.

CHAPTER V.

The Third Period.

It was stated in the opening chapter that the third of the periods into which we arbitrarily divide the history of Japanese mathematics was less than a century in duration, extending from about 1600 to about 1675. The first of these dates is selected as marking approximately the beginning of the activity of Mōri Kambei Shigeyoshi, who was mentioned in Chapter III, and the last as marking that of Seki. It was an era of intellectual awakening in Japan, of the welcoming of Chinese ideas, and of the encouragement of native effort. Of the work of Mōri we have already spoken, because he had so much to do with making known, and possibly improving, the *soroban*. It now remains to speak of his pupils, and first of Yoshida.

Yoshida Shichibei Kōyū, or Mitsuyoshi, was born at Saga, near Kyōto, in 1598, as we are told in Kawakita's manuscript, the *Honchō Sūgaku Shiryō*. He belonged to an ancient family that had contributed not a few illustrious names to the history of the country. Yoshida Sōkei, for example, who died in 1572, was well known in medicine, and had twice made a journey to China in search of information, once with a Buddhist bonze[1] in 1539, and again in 1547. His son Kōkō, (1554—1616), was a noted engineer, and is known for his work in improving navigation on the Fujikawa and other rivers that had been too dangerous for the passage of boats. Kōkō's son Soan was, like his father, well known for his learning and for his engineering skill.[2] Yoshida Kōyū, the mathematician, was a

---

[1] Priest. The name is a Portuguese corruption of a Japanese term.

[2] See the *Sentetsu Sōdan Zoku-hen*, 1884, Book I.

grandson, on his mother's side, of Yoshida Kōkō.[1] He was also related in another way to the Yoshida family, being the eldest son of Yoshida Shūan, who was the great-grandson of Sōkei's father, Sōchū.

Yoshida, as we shall now call him, early manifested a taste for mathematics, going as a youth to Kyōto that he might study under the renowned Mōri. His ignorance of Chinese was a serious handicap, however, and his progress was a disappointment. He thereupon set to work to learn the language, studying under the guidance of his relative Yoshida Soan, and in due time became so proficient that he was able to read the *Suan-fa T'ung-tsong* of Ch'êng Tai-wei.[2] His progress in mathematics then became so rapid that it is related[3] that he soon distanced his master, so that Mōri himself was glad to become his pupil. Yoshida also continued to excel in Chinese, so that, whereas Mōri knew the language only indifferently, his quondam pupil became master of the entire mathematical literature.

Mōri's works were the earliest native Japanese books on mathematics of which we have any record, but they seem to be irretrievably lost. It is therefore to Yoshida that we look as the author of the oldest Japanese work on mathematics extant. This work was written in 1627 and is entitled *Jinkō-ki*. The name is interesting, the Chinese ideogram *jin* meaning (among other things) a small number, *kō* meaning a large number, and *ki* a treatise, so that the title signifies a treatise on numbers from the greatest to the least. Yoshida tells us in the preface that it was selected for him by one Genkō, a Buddhist priest, and it is typical of the condensed expressions of the Japanese.

The work relates chiefly to the arithmetical operations as performed on the *soroban*, including square and cube root, but it also has some interesting applications and it gives 3.16 for

---

1 Endō, Book I, p. 35.

2 Which had appeared in 1593. See p. 34.

3 By Kawakita in the *Honchō Sugaku Shiryō*.

the value of π. It is based largely upon the *Suan-fa T'ung-tsong* already described, and the preface states that it originally consisted of eighteen books. Only three books have come down to us, however, and indeed we are assured that only three were ever printed.[1] This was the first treatise on mathematics ever printed in Japan, or at least the first of any importance.[2] It appeared in 1627[3] and was immediately received with great enthusiasm. Even during Yoshida's life a number of editions appeared,[4] and the name *Jinkō-ki* was used so often after his death, by other authors, that it became a synonym for arithmetic, as algorismus did in Europe in the late Middle Ages.[5] Indeed it is hardly too much to compare the celebrity of the *Jinkō-ki* in Japan with that of the arithmetic of Nicomachus in the late Greek civilization. Yoshida also wrote on the calendar, but these works[6] were not so well known.

So great was the fame of Yoshida that he was called to the court of Hosokawa, the feudal lord of Higo, that he might instruct his patron in the art of numbers Here he resided for a time, and at his lord's death, in 1641, he returned to his native place and gathered about him a large number of pupils, even as Mōri had done before him. In his declining years an affection of the eyes, which had troubled him from his youth, became more serious, and finally resulted in the affliction of

---

[1] By the bonze Genkō who wrote the preface, and by Yoshida himself at the end of the 1634 edition.

[2] Mr. Endō has shown the authors the copy of the edition of 1634 in the library of the Tōkyō Academy and has assured us that the edition of 1627 was the first Japanese mathematical work of any importance. There is a tradition that Mōri's *Kijo Ranjo* was also printed.

[3] That is, the 4th year of Kwan-ei.

[4] As in 1634, 1641, and 1669, all edited by Yoshida. There were several pirated editions. See Muramatsu's *Sanso* of 1663, Book III; Endō, Book I, pp. 58, 59, 84 etc.

[5] Compare the German expression "Nach Adam Riese", the English "According to Cocker", the early American "According to Daboll", and the French word *Barême*.

[6] For example, the *Wakan Gō-un* and the *Koreki Benran*.

total blindness,—the fate of Saunderson and of Euler as well. He died in 1672 at the age of seventy-four.[1]

The immediate effect of the work of Mōri and Yoshida was a great awakening of interest in computation and mensuration. In 1630 the Shogun established the *Kōbun-in,* a public school of arts and sciences. Unfortunately, however, mathematics found no place in the curriculum, remaining in the hands of private teachers, as in the days of the German Rechenmeister. Nevertheless the science progressed in a vigorous manner and numerous books were published upon the subject. Yoshida had appended to one of the later editions of his *Jinkō-ki* a number of problems with the proposal that his successors solve them. These provoked a great deal of discussion and interest, and led other writers to follow the same plan, thus leading to the so-called *idai shōtō,*[2] "mathematical problems proposed for solution and solved in subsequent works". This scheme was so popular that it continued until 1813, appearing for the last time in the *Sangaku Kōchi* of Ishiguro Shin-yū.

The particular edition of Yoshida's *Jinkō-ki* in which these problems appeared is not extant, but the problems are known through their treatment by later writers, and some of them will be given when we come to speak of the work of Isomura.

The second of Mōri's "three honorable scholars" mentioned in Chapter III was Imamura Chishō, and twelve years after the appearance of the *Jinkō-ki,* that is in 1639, he published a treatise entitled, *Jugai-roku.*[3] Yoshida's work had appeared in Japanese, although it followed the Chinese style, but Imamura wrote in classical Chinese. Beginning with a treatment of the *soroban,* he does not confine himself to arithmetic, as Yoshida had done, but proceeds to apply his number work to the calculations of areas and volumes, as in the case of the

---

[1] C. KAWAKITA, *Honchō Sugaku Shiryō*; ENDŌ, Book I, p. 84.

[2] A term used by later scholars.

[3] Mr. Endō has shown the authors a copy of Andō's commentary in the library of the Academy of Science at Tōkyō, and Dr. K. Kano has a copy of the original at present in his valuable library. At the end of the work the author states that only a hundred copies were printed.

circle, the sphere, and the cone. While Yoshida had taken 3.16 for the value of π, Imamura takes 3.162. Andō Yūyeki of Kyōto refers to this in his *Jugai-roku Kana-shō*, printed in 1660, as obtained by extracting the square root of 10. If this is true, Yoshida obtained his in the same way, the square root of 10 having long been a common value for π in India and Arabia, as well as in China. Liu Hui's commentary on the "Nine Sections" asserts that the first Chinese author to use this value was Chang Hêng, 78—139 A. D. It was also used by Ch'ên Huo in the eleventh century, and by Ch'in Chiu-shao in his *Su-shu Chiu-chang* of 1247.[1] Some Chinese writers even in the present dynasty have used it, and it was very likely brought from that country to Japan. It is of interest to note that lumbermen and carpenters in certain parts of Japan use this value at the present time.

Imamura gives as a rule for finding the area of a circle that the product of the circumference by the diameter should be divided by 4. The volume of the sphere with diameter unity is given as 0.51, which does not fit his value of π as closely as might have been expected. He also gives a number of problems about the lengths of chords, and writes extensively upon the *Kaku-jutsu* or "polygonal theory",—calculations relating to the regular polygons from the triangle to the decagon. This theory attracted considerable attention on the part of his successors and added much to Imamura's reputation.[2] This treatise was translated into Japanese and a commentary was added by Imamura's pupil, Andō Yūyeki, in 1660.

The year following the appearance of the original edition Imamura published the *Inki Sanka* (1640), a little work on the *soroban*, written in verse. The idea was that in this way the rules could the more easily be memorized, an idea as old as civilization. The Hindus had followed the same plan many

---

1 MIKAMI, Y., *On the development of the Chinese mathematics* (in Japanese), in the *Journal of the Tōkyō Physics School*, No. 203, p. 450; *Mathematical papers from the Far East*, Leipzig, 1910, p. 5.

2 ENDŌ, Book I, pp. 59, 60.

centuries earlier, and a generation or so before Imamura wrote it was being followed by the arithmetic writers of England.

The third of the *San-shi* of Mōri was Takahara Kisshu, also known as Yoshitane.[1] While he contributed nothing in the way of a published work, he was a great teacher and numbered among his pupils some of the best mathematicians of his time.

During this period of activity numerous writers of prominence appeared, particularly on the *soroban* and on mensuration. Among these writers a few deserve a brief mention at this time. Tawara Kamei wrote his *Shinkan Sampō-ki* in 1652,

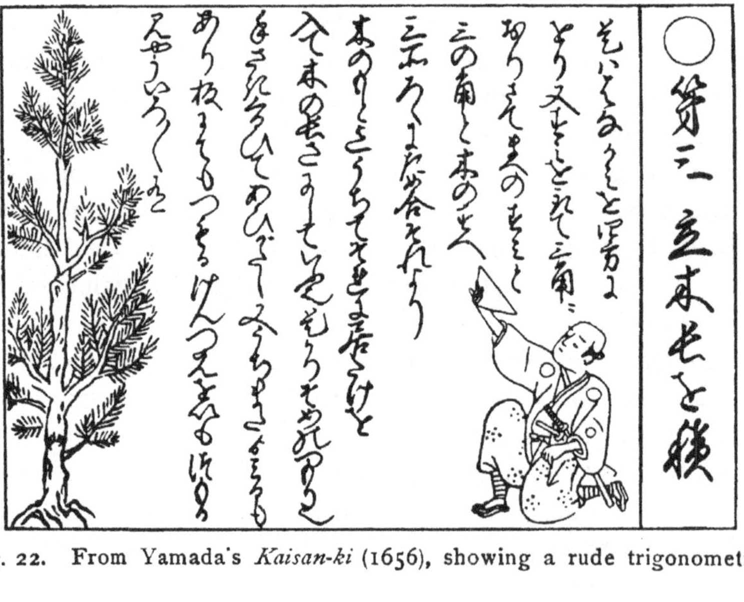

Fig. 22. From Yamada's *Kaisan-ki* (1656), showing a rude trigonometry.

and Yenami Washō his *Sanryō-roku* in the following year. In 1656 Yamada Jūsei published the *Kaisan-ki* (Fig. 22) which was very widely read, and the title of which was adopted, with various prefixes, by several later writers. The following year (1657) saw the publication of Hatsusaka's *Yempō Shikan-ki* and Shibamura's *Kakuchi Sansho.* A year later (1658) appeared Nakamura's *Shikaku Mondō,* followed in 1660 by Isomura's *Ketsugi-shō,* in 1663 by Muramatsu's *Sanso,* in 1664

[1] The names are synonyms.

by Nozawa Teichō's *Dōkai-shō,* and in 1666 by Satō's *Kongenki.* These are little more than names to Western readers, and yet they go to show the activity that was manifest in the field of elementary mathematics, largely as the result of the labors of Mōri and of Yoshida. The works themselves were by no means commercial arithmetics, for they perfected little by little the subject of mensuration, the method of approximating the value of π, and the treatment of the regular polygons, besides offering a considerable insight into the nature of magic squares and magic circles. To these books we are indebted for our knowledge of the work of this period, and particularly to the *Kaisan-ki* (1656), the *Shikaku-Mondō* (1658), and the *Ketsugi-shō,* (1660).

The last mentioned work, the *Ketsugi-shō,* was written by a pupil of Takahara Kisshu,[1] who was one of the *San-shi* of Mōri. His name was Isomura[2] Kittoku, and he was a native of Nihommatsu in the north-eastern part of Japan. Isomura's *Ketsugi-shō*[3] appeared in five books in 1660, and was again published in 1684 with notes. We know little of his life, but he must have been very old when the second edition of his work appeared for he tells us in the preface that at that time he could hardly hold a *soroban* or the *sangi.*

Two features of the *Ketsugi-shō* deserve mention,—Isomura's statement of the Yoshida problems (including an approach to integration, as seen in Fig. 23) and similar ones of his own, and his treatment of magic squares and circles. Each of these throws a flood of light upon the nature of the mathematics of Japan in its Renaissance period, just preceding the advent of the greatest of her mathematicians, Seki, and each is therefore

---

1 OZAWA, *Sanka Furyaku,* "Brief Lineage of Mathematicians", manuscript of 1801.

2 ENDŌ gives it as ISOMURA, Book I, pp. 65, 67, and Book II, p. 20 etc., and in this he was at first followed by HAYASHI, *History,* part I, p. 33, although the latter soon after discovered that IWAMURA was the better form. HAYASHI gives the personal name as Yoshinori.

3 Or *Sampō-ketsugi-shō.*

worthy of our attention. Of the Yoshida problems the following are types:[1]

"There is a log of precious wood 18 feet[2] long, whose bases are 5 feet and $2\frac{1}{2}$ feet in circumference. ... Into what lengths should it be cut to trisect the volume?"

"There have been excavated 560 measures of earth which are to be used for the base of a building.[3] The base is to be 30 measures square and 9 measures high. Required the size of the upper base."

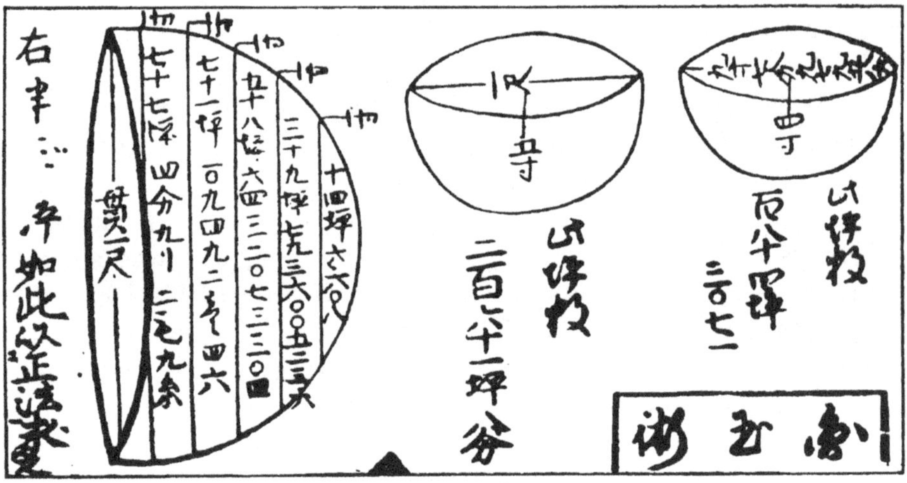

Fig. 23. From the second (1684) edition of Isomura's *Ketsugi-shō*.

"There is a mound of earth in the form of the frustum of a circular cone. The circumferences of the bases are 40 measures and 120 measures, and the mound is 6 measures high. If 1200 measures of earth are taken evenly off the top, what will then be the height?"

"A circular piece of land 100 measures in diameter is to be divided among three persons so that they shall receive 2900,

1 The *Ketsugi-shō* of 1660, Book 4.

2 In the original "3 measures".

3 That is, for a mound in the form of a frustum of a square pyramid.

2500, and 2500 measures respectively.[1] Required the lengths of the chords and the altitudes of the segments."

The rest of the problems relate to the triangle and to linear simultaneous equations of the kind found in such works as the "Nine Sections", the *Suan-fa T'ung-tsong*, and the *Suan-hsiao Chi-mêng*. The last of the problems given above is solved by Isomura as follows:—

"Divide 7900 measures,[2] the total area, by 2900 measures of the northern segment, the result being 2724.[3] Double this result and we have 5448. Divide the square of the diameter, 100 measures, by 5448 and the result is 1835.554[4] measures. The square root of this is 42.85 measures. Subtract this from half the diameter and we have 7.15 measures. Multiply the 42.85 by this and we have 306.4 measures. We now multiply by a certain constant for the square and the circle, and divide by the diameter and we have 3.45 measures. Subtract this from 42.85 measures and we have 39.4 measures for the height of the northern segment..."

Following Yoshida's example, Isomura gives a series of problems for solution, a hundred in number, placing them in his fifth book. A few of these will show the status of mathematics at the time of Isomura:

"From a point in a triangle lines are drawn to the vertices. Given the lengths of these lines and of two sides of the triangle, to find the length of the third side of the triangle." (No. 28.)

"A string 62.5 feet long is laid out so as to form Seimei's Seal.[5] Required the length of the side of the regular pentagon in the center." (No. 38.)

"A string is coiled so as first to form a circle 0.05 feet in diameter, and [then so that the coils shall] always keep 0.05 feet apart, and the coil finally measures 125 feet in diameter.

---

[1] By drawing two parallel chords.

[2] It would have been 7854 if he had taken $\pi = 3.1416$.

[3] I. e., 2.724+.

[4] Where we now introduce the fraction for clearness.

[5] Abe no Seimei was a famous astrologer who died in 1005. His seal was the regular pentagonal star, the badge of the Pythagorean brotherhood.

What is the length of the string?" (No. 39.) The reading of this problem is not clear, but Isomura seems to mean that a spiral of Archimedes is to be formed coiled about an inner circle, and finally closing in an outer circle. The curve has attracted a good deal of attention in Japan.

"There is a log 18 feet long, the diameter of the extremities being 1 foot and 2.6 feet respectively. This is wound spirally with a string 75 feet long, the coils being 2.5 feet apart. How many times does the string go around it?" (No. 41.)

"The bases of a frustum of a circular cone have for their respective diameters 50 measures and 120 measures, and the height of the frustum is 11 measures. Required to trisect the volume by planes perpendicular to the base." (No. 44.)

"The bases of a frustum of a circular cone have for their respective diameters 120 and 250 measures, and the height of the frustum is 25 measures. The frustum is to be cut obliquely. Required the perimeter of the section." (No. 45.) Presumably the cutting plane is to be tangent to both bases, thus forming a complete ellipse, a figure frequently seen in Japanese works.

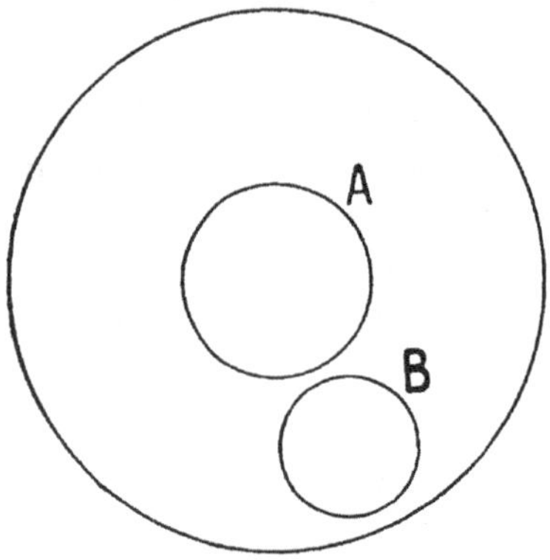

"In a circle 3 feet in diameter 9 other circles are to be placed, each being 0.2 of a foot from every other and from the large circle. Required the diameter of the larger circle in the center, and of the smaller circles surrounding it." (No. 60.) This requires us to place a circle A in the center, arranging eight smaller circles B about it so as to satisfy the conditions.

"If 19 equal circles are described outside a given circle that has a circumference of 12 feet, so as to be tangent to the given circle and to each other; and if 19 others are similarly described within the given circle, what will be the diameters of the circles in these two groups?" (No. 61.)

"To find the length of the minor axis of an ellipse whose area is 748.940625, and whose major axis is 38 measures." (No. 84.)

"To find one axis of an ellipsoid of revolution, the other axis being 1.8 feet, and the volume being 2422, the unit of volume being a cube whose edge is 0.1 of a foot." (N. 85.) Here the major axis is supposed to be the axis of revolution.

Isomura was also interested in magic squares, and these forms were evidently the object of much study in his later years, since the 1684 edition of his *Ketsugi-shō* contains considerable material relating to the subject. In the first edition (1660) there appear both odd and even-celled squares. The following types suffice to illustrate the work.[1]

40	38	2	6	1	42	46
41	20	17	37	19	32	9
3	16	26	21	28	34	47
39	36	27	25	23	14	11
43	35	22	29	24	15	7
5	18	33	13	31	30	45
4	12	48	44	49	8	10

55	4	2	62	64	60	6	7
51	20	22	17	50	42	44	14
9	49	40	28	25	37	16	56
12	46	29	31	34	36	19	53
13	18	35	33	32	30	47	52
54	41	26	38	39	27	24	11
8	21	43	48	15	23	45	57
58	61	63	3	1	5	59	10

[1] It should be said that the history of the magic square has never adequately been treated. Such squares seem to have originated in China and to have spread throughout the Orient in early times. They are not found in the classical period in Europe, but were not uncommon during and after the 12th century. They are used as amulets in certain parts of the world, and have always been looked upon as having a cabalistic meaning. For a study of the subject from the modern standpoint see ANDREWS, W. S., *Magic Squares*, Chicago, 1907, and subsequent articles in *The Open Court*.

51	46	53	6	1	8	69	64	71
52	50	48	7	5	3	70	68	66
47	54	49	2	9	4	65	72	67
60	55	62	42	37	44	24	19	26
61	59	57	43	41	39	25	23	21
56	63	58	38	45	40	20	27	22
15	10	17	78	73	80	33	28	35
16	14	12	79	77	75	34	32	30
11	18	13	74	81	76	29	36	31

92	91	15	89	4	84	14	99	11	6
13	73	22	20	80	83	78	24	25	88
85	69	38	40	35	68	60	62	32	16
3	27	67	58	46	43	55	34	74	98
96	30	64	47	49	52	54	37	71	5
8	31	36	53	51	50	48	65	70	93
18	72	59	44	56	57	45	42	29	83
94	26	39	61	66	33	41	63	75	7
1	76	79	81	21	19	23	77	28	100
95	10	86	12	97	17	87	2	90	9

In the last (1684) edition he gives a number of new arrangements, including the following:

4	9	5	16
14	7	11	2
15	6	10	3
1	12	8	13

5	23	16	4	25
15	14	7	18	11
24	17	13	9	2
20	8	19	12	6
1	3	10	22	21

10	8	35	33	24	1
19	26	17	15	6	28
5	12	30	34	16	14
23	21	3	7	25	32
18	31	22	20	11	9
36	13	4	2	29	27

Isomura did also a good deal of work on magic circles, the following appearing in his 1660 edition:

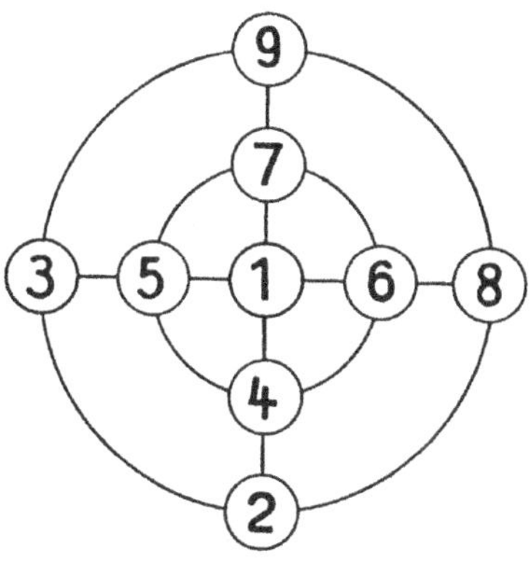

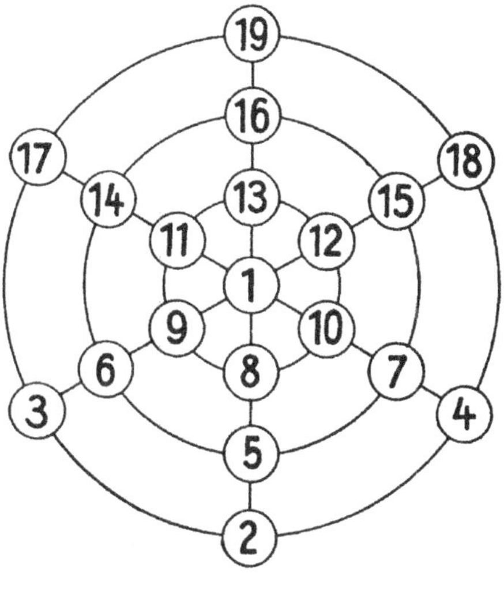

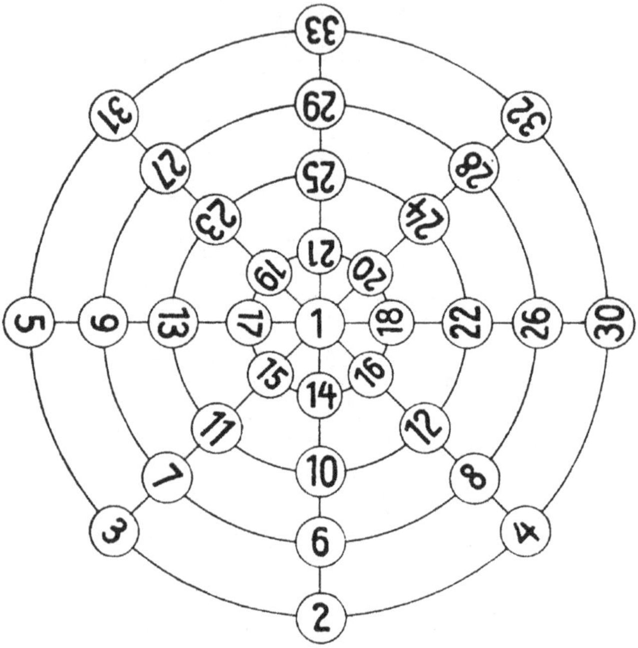
1
2
3
4
5
6
7
8
9
10
11
12
13
14
15
16
17
18
19
20
21
22
23
24
25
26
27
28
29
30
31
32
33

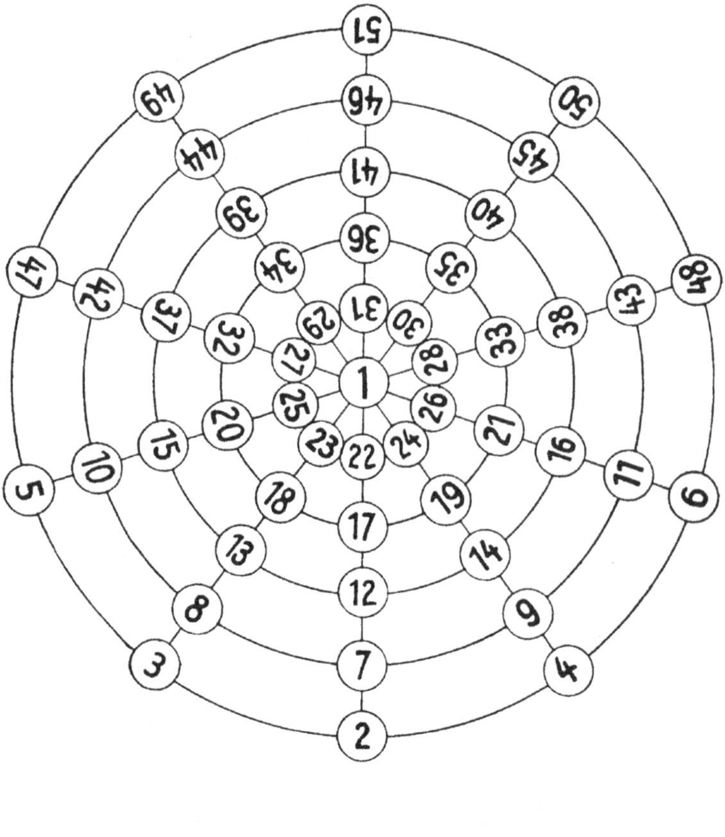
1
2
3
4
5
6
7
8
9
10
11
12
13
14
15
16
17
18
19
20
21
22
23
24
25
26
27
28
29
30
31
32
33
34
35
36
37
38
39
40
41
42
43
44
45
46
47
48
49
50
51

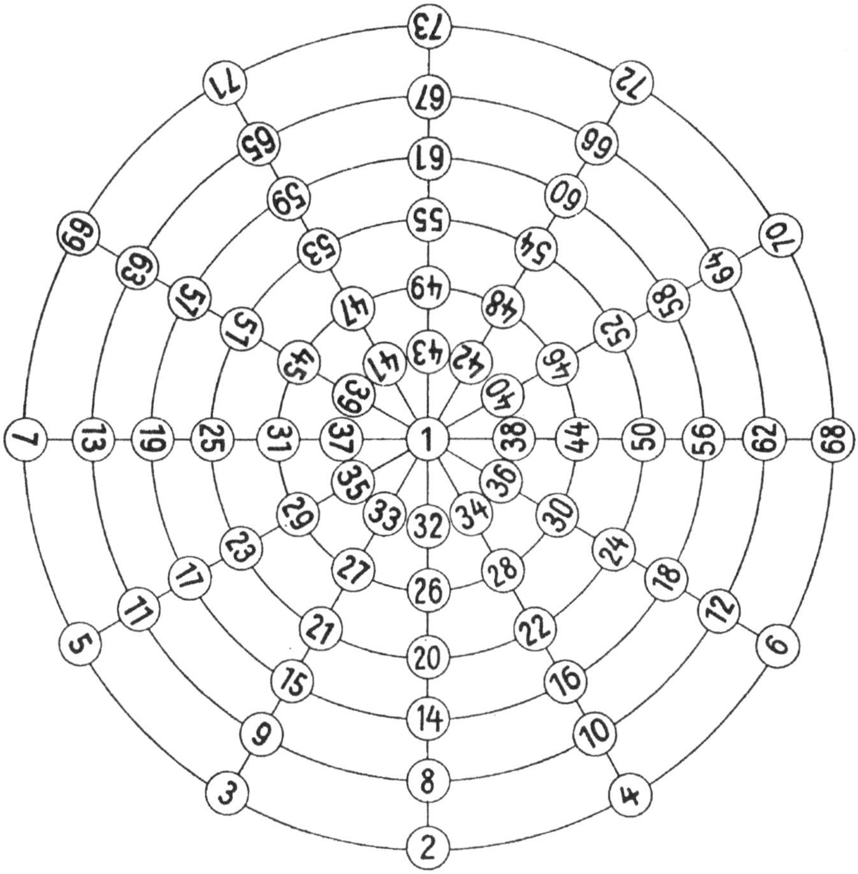

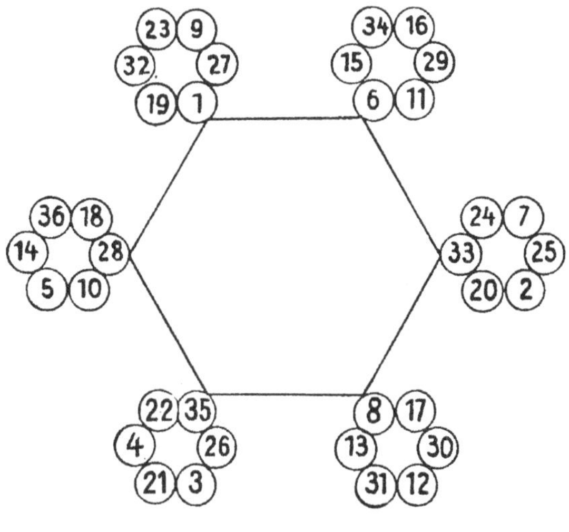

In the 1684 edition of his *Ketsugi-shō* he gives what he calls sets of magic wheels. Here, and on pages 74 and 75, the sums in the minor circles are constant.

Isomura's method[1] of finding the area of the circle is as

[1] 1660 edition of the *Ketsugi-shō*, Book III.

follows: Take a circle of diameter 10 units, and divide the circumference into parts whose lengths are each a unit. It will then be found that there are 31 of these equal arcs, with a smaller arc of length 0.62. Join the points of division to the center, thus making a series of triangular shaped figures. By

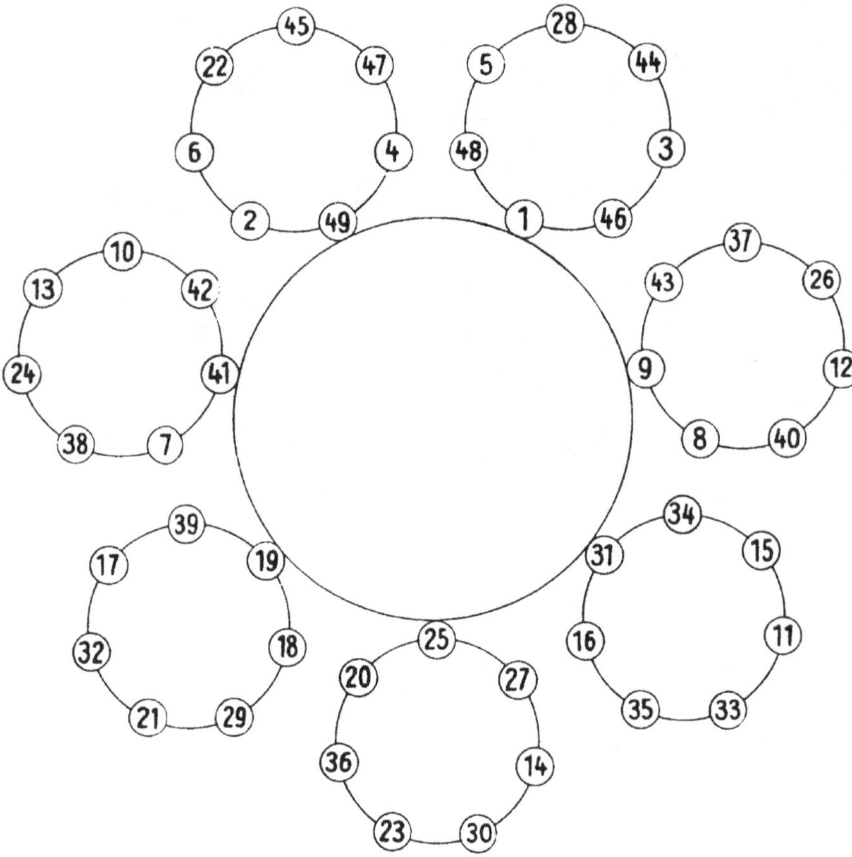

dove-tailing these triangles together we can form a rectangular shaped figure whose length is 15.81, and whose width is 5, so that the area equals 5×15.81, or 79.05. Hence, in modern notation, $\frac{\pi}{4} \times$ diameter is the area.

In the 1660 edition of the *Ketsugi-shō* he gives the surface of a sphere as one-fourth the square of its circumference, which would make it $\pi^2 r^2$ instead of $4\pi r^2$. In the 1684 edition,[1] however, he says that this is incorrect, although he asserts that it had been stated by Mōri, Yoshida, Imamura, Takahara, Hiraga, Shimada, and others. It seems that the rule had been derived from considering the surface of the sphere as if it were

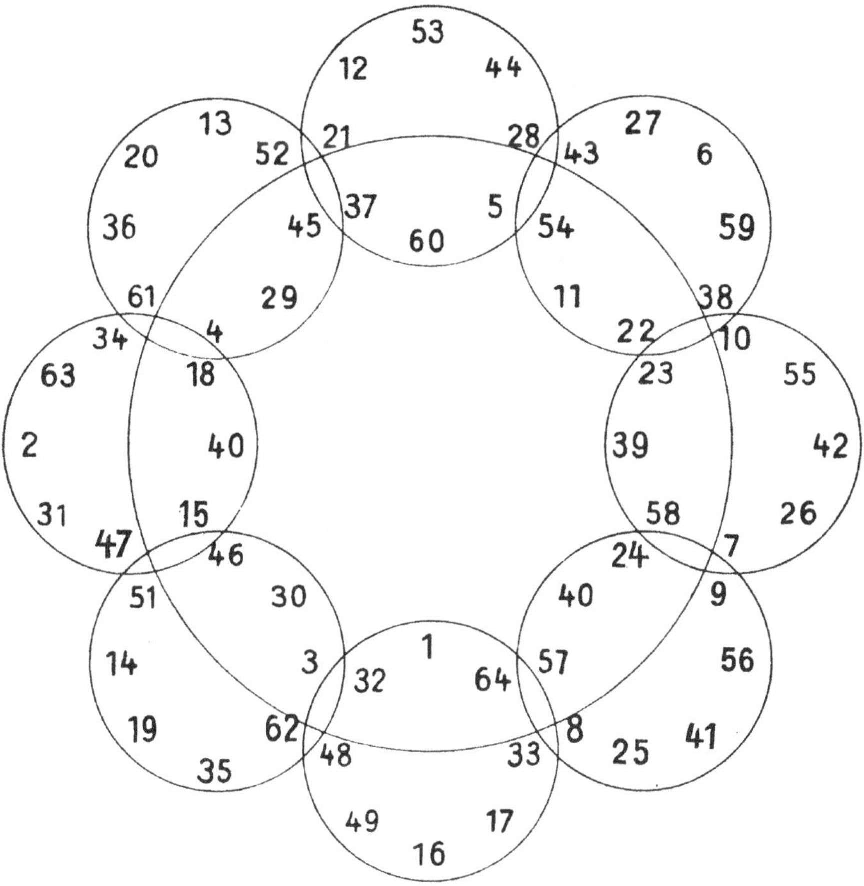

the skin of an orange that could be removed and cut into triangular forms and fitted together in the same manner as the sectors of a circle. The error arose from not considering the curvature of the surface. To rectify the error Isomura

---

[1] Book IV, note.

took two concentric spheres with diameters 10 and 10.0002. He then took the differences of their volumes and divided this by 0.0001, the thickness of the rind that lay between the two surfaces. This gave for the spherical surface 314.160000041888, from which he deduced the formula, $s = \frac{6v}{d} = \pi d^2$. This ingenious process of finding s, which of course presupposes the ability to find the volume of a sphere, has since been employed by several writers.[1]

It should be mentioned, before leaving the works of Isomura, that the 1684 edition of the *Ketsugi-shō* contains a few notes in which an attempt is made to solve some simultaneous linear equations by the method of the "Celestial element" already described. The author states, however, that he does not favor this method, since it seems to fetter the mind, the older arithmetical methods being preferable.

Isomura seems not to have placed in his writings all of his knowledge of such subjects as the circle, for he distinctly states that one must be personally instructed in regard to some of these measures. Possibly he was desirous of keeping this knowledge a secret, in the same way that Tartaglia wished to keep his solution of the cubic. Indeed, there is a 19th century manuscript that is anonymous, although probably written by Furukawa Ken, bearing the title *Sanwa Zuihitsu* (Miscellany about Mathematical Subjects), in which it is related that Isomura possessed a secret book upon the mensuration of the circle, and in particular upon the circular arc. It is said that this was later owned by Watanabe Manzō Kazu, one of Aida Ammei's pupils, and a retainer of the Lord of Nihommatsu, where Isomura one time dwelt. The writer of the *Sanwa Zuihitsu* asserts that he saw the book in 1811, during a visit at his home by Watanabe, and that he made a copy of it at that time. He says that the methods were not modern and that they contained fallacies, but that the explanations were

---

1 It is given in Takebe Kenkō's manuscript work, the *Fukyū Tetsujutsu* of 1722, in an anonymous manuscript entitled *Kigenkai*, and in a work of the 19th century by Wada Nei.

minute. The title of the work was *Koshigen Yensetsu Hompō*, and it was dated the 15th day of the 3d month of 1679.

Next in rank to Isomura, in this period, was Muramatsu Kudayū Mosei.[1] He was a pupil of Hiraga Yasuhide, a distinguished teacher but not a writer, who served under the feudal Lord of Mito, meeting with a tragic death in 1683.[2]

Muramatsu was a retainer of Asano, Lord of Akō, whose forced suicide caused the heroic deed of the "Forty-seven Rōnins" so familiar to readers of Japanese annals. Muramatsu is sometimes recorded as one of the honored "Forty-seven", but it was his adopted son, Kihei, and Kihei's son, who were among the number.[3] As to Muramatsu himself, he died at an advanced age after a life of great activity in his chosen field.

In 1663 Muramatsu began the publication of a work in five books, entitled the *Sanso*.[4] In this he treats chiefly of arithmetic and mensuration, following in part the Chinese work, *Suan-hsiao Chi-mêng*, written by Chu Chi-chieh, as mentioned on page 48, but he fails to introduce the method of the "Celestial element". The most noteworthy part of his work relates to the study of polygons[5] and to the mensuration of the circle.[6]

Taking the radius of the circumscribed circle as 5, he calculates the sides of the regular polygons as follows:

No. of sides.	*Length of side.*	*No. of sides.*	*Length of side.*
5	5.8778	11	2.801586
6	5	12	2.5875
7	4.3506	13	2.393
8	3.82682	14	2.22678
9	3.4102	15	2.07953
10	3.0876	16	1.95093

---

1 Not Matsumura, as given by ENDŌ. The name Mosei appears as Shigekiyo in his *Mantoku Jinkō-ki* (1665).

2 See the *Stories told by Araki*.

3 AOYAMA, *Lives of the Forty-seven Loyal Men* (in Japanese).

4 The last book bears the date 1684, and may not have appeared earlier.

5 Book 2. 6 Book 4.

To calculate the circumference Muramatsu begins with an inscribed square whose diagonal is unity. He then doubles the number of sides, forming a regular octagon, the diameter of the circumscribed circle being one. He continues to double the number of sides until a regular inscribed polygon of 3278 sides is reached. He computes the perimeters of these sides in order, by applying the Pythagorean Theorem, with the following results:

No. of sides.	*Perimeter.*
2^3	3.06146745892071817384
2^4	3.12144515225805237021 3
2^5	3.13654849054593934785 3
2^6	3.1403311569547 53
2^7	3.14127925093277291340 16
2^8	3.14151380114430112844 8
2^9	3.14157294036709143516 2
2^{10}	3.14158772527715976659
2^{11}	3.14159142151118673329 6
2^{12}	3.14159234557010467614 72
2^{13}	3.14159257658486051086 81
2^{14}	3.14159263433855298
2^{15}	3.14159264877769886924 8

Having reached this point, Muramatsu proceeded to compare the various Chinese values of π, and stated his conclusion that 3.14 should be taken, unaware of the fact that he had found the first 8 figures correctly.[1]

Muramatsu gives a brief statement as to his method of finding the volume of a sphere, but does not enter into details.[2] He takes 10 as the diameter, and by means of parallel planes he cuts the sphere into 100 segments of equal altitude. He then assumes that each of these segments is a cylinder, either with the greater of the two bases as its base, or with the lesser one. If he takes the greater base, the sum of the vol-

[1] Endō, Book I, p. 70.

[2] The *Sanso*, Book 5.

umes is 562.5 cubic units; but if he takes the lesser one this sum is only 493.04 cubic units. He then says that the volume of the sphere lies between these limits, and he assumes, without,

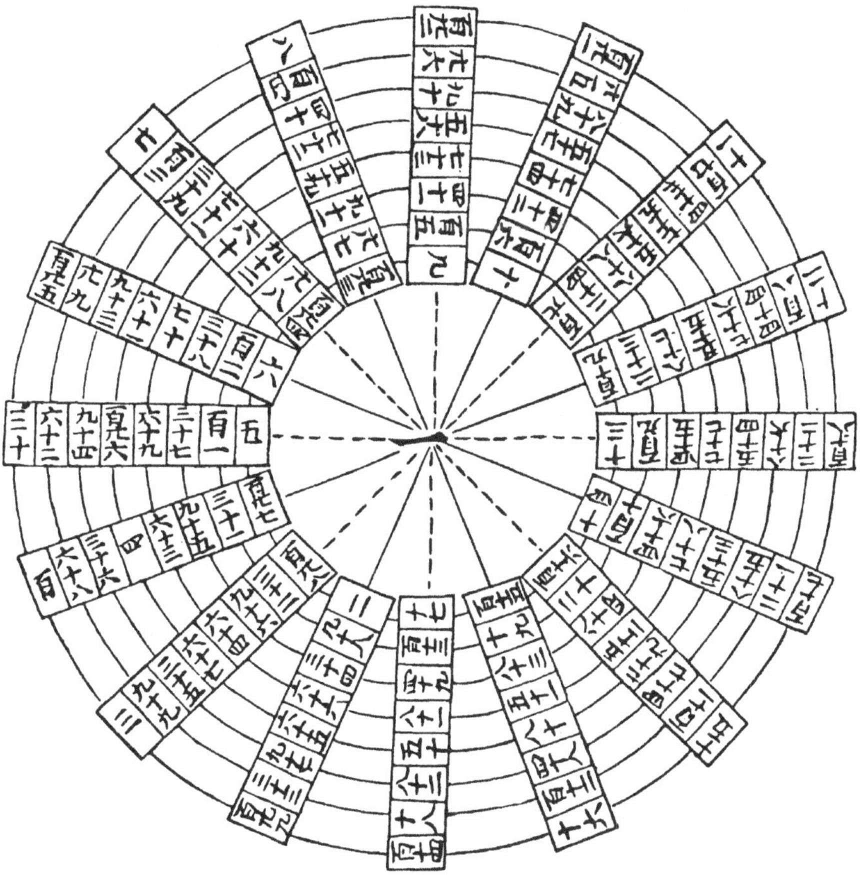

Fig. 24. Magic circle, from Muramatsu Kudayū Mosei's *Mantoku Jinkō-ki* (1665).

stating his reasons, that it is 524, which is somewhat less than either their arithmetic (527) or their geometric (526.6) mean,[1] and which is equivalent to taking π as 3.144.

Muramatsu was also interested in magic squares[2] and magic

[1] ENDŌ thinks that he may have reached this value by cutting the sphere into 200, 400 or some other number of equal parts. *History,* Book I, p. 71.

[2] His *rakusho* (afterwards called *hōjin*) problems.

circles.[1] One of his magic squares has 19^2 cells, as did one published by Nozawa Teichō in the following year.[2] One of his magic circles, in which 129 numbers are used, is shown in Fig. 24 on page 79. With the numbers expressed in Arabic numerals it is as follows:

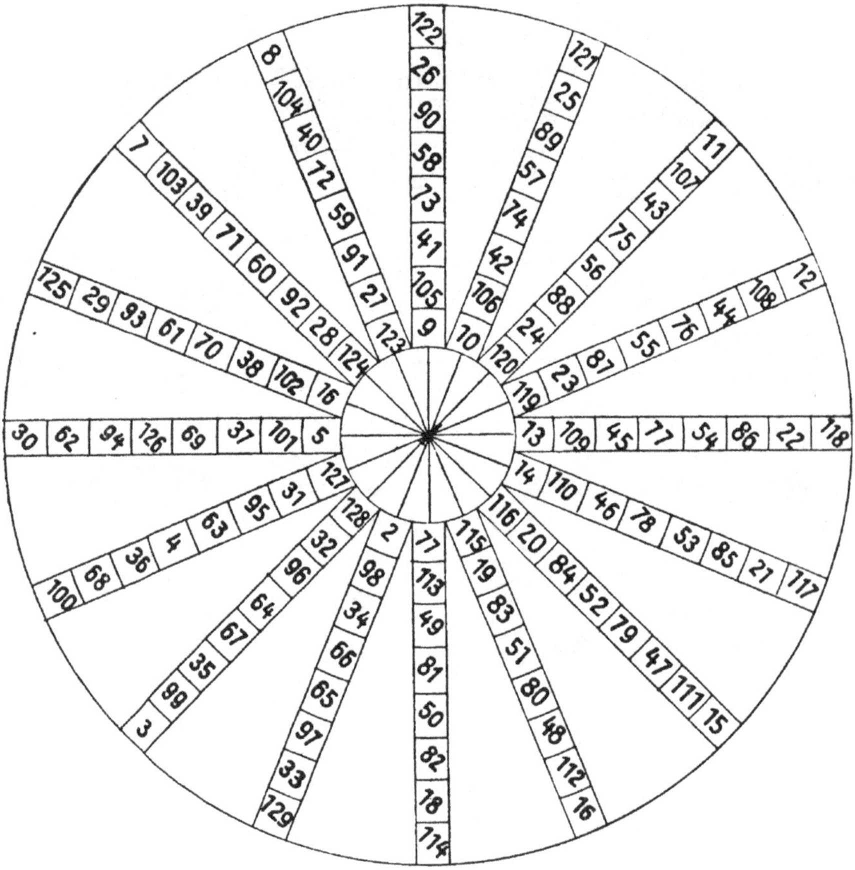

In Muramatsu's work also appears a variant of the famous old Josephus problem, as it is often called in the West, a problem that had already appeared in the *Jinkō-ki* of Yoshida.

1 His *ensan* problems. *Sanso*, Book 2.

2 In his *Dōkai-shō* of 1664.

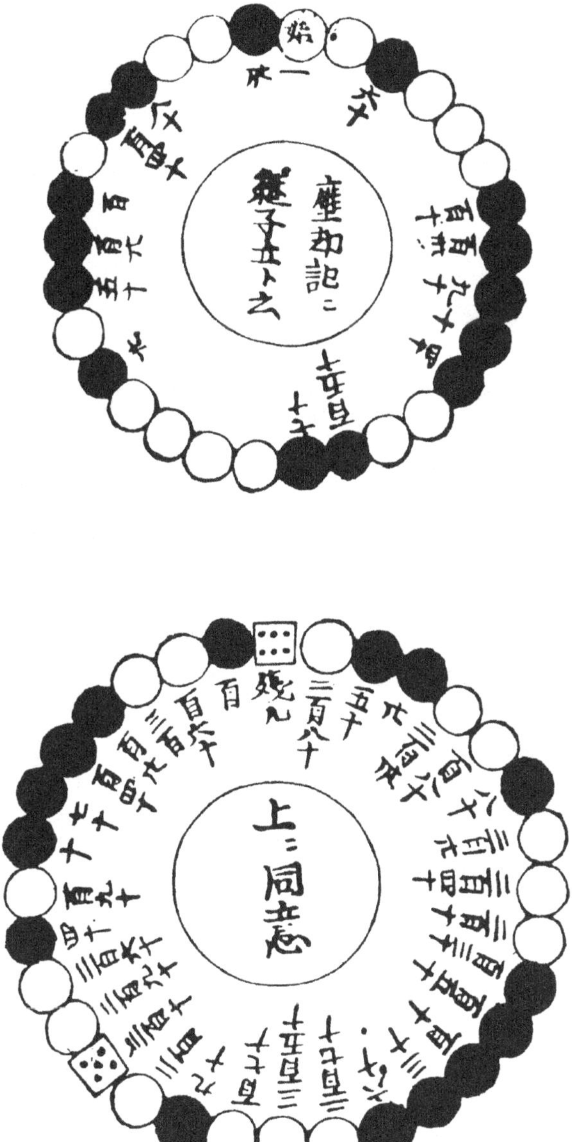

Fig. 25. The Josephus problem, from Muramatsu Kudayū Mosei's *Mantoku Jinkō-ri* (1665).

As given by Seki, a little later, it is as follows: "Once upon a time there lived a wealthy farmer who had thirty children, half being born of his first wife and half of his second one. The latter wished a favorite son to inherit all the property, and accordingly she asked him one day, saying: Would it not be well to arrange our thirty children on a circle, calling

Fig. 26. The Josephus problem, from Miyake Kenryū's *Shojutsu Sangaku Zuye* (1795 edition).

one of them the first and counting out every tenth one until there should remain only one, who should be called the heir. The husband assenting, the wife arranged the children as shown in the figure[1]. The counting then began as shown and resulted in the elimination of fourteen step-children at once, leaving only one. Thereupon the wife, feeling confident of her success,

[1] The step children are represented by dark circles, and her own children by light ones. In the old manuscripts the latter are colored red.

said: Now that the elimination has proceeded to this stage, let us reverse the order, beginning with the child I choose. The husband agreed again, and the counting proceeded in the reverse order, with the unexpected result that all of the second wife's children were stricken out and there remained only the step-child, and accordingly he inherited the property." The original is shown in Fig. 25, and an interesting illustration from Miyake's work of 1795 in Fig. 26, but the following diagram will assist the reader:

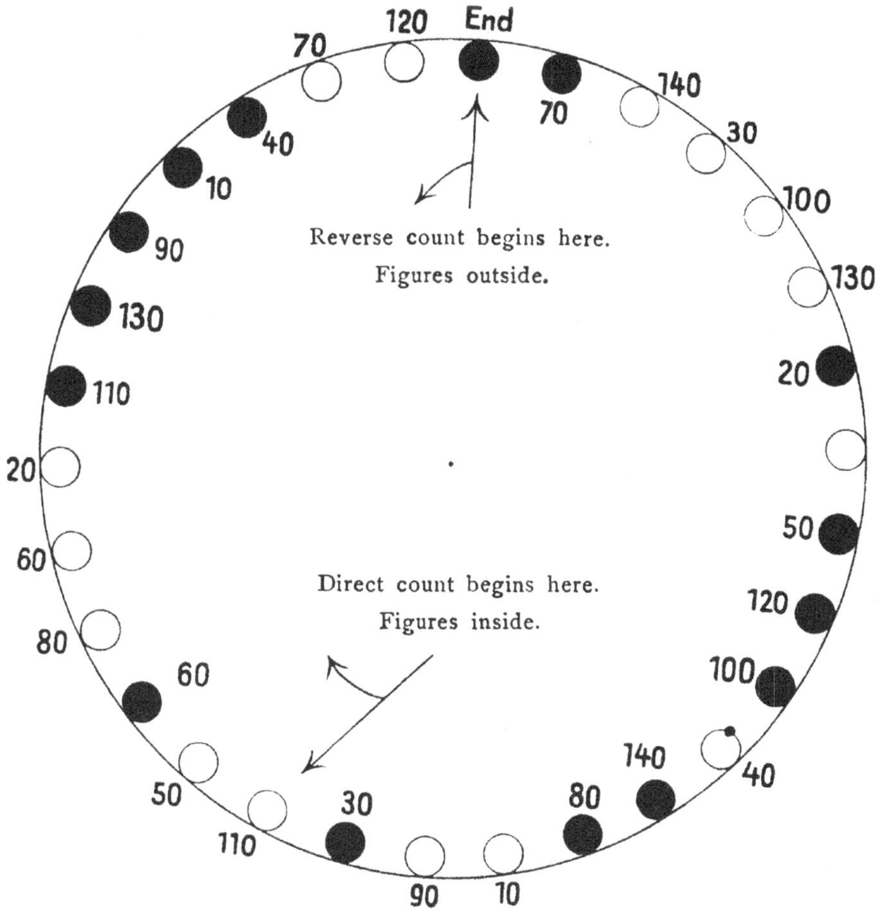

Perhaps it is more in accord with oriental than with occidental nature that the interesting agreement should have remained in force, with the result that the heir should have been a step-son of the wife who planned the arrangement. Seki also gave the problem, having obtained it from the *Jinkō-ki* of Yoshida, although he mentions only the fact that it is an old tradition. Possibly it was one of Michinori's problems in the twelfth century, but whether it started in the East and made its way to the West, or vice versa, we do not know. The earliest definite trace of the analogous problem in Europe is in the Codex Einsidelensis, early in the tenth century, although a Latin work of Roman times[1] attributes it to Flavius Josephus. It is also mentioned in an eleventh century manuscript in Munich and in the *Ta'hbula* of Rabbi Abraham ben Ezra (d. 1067). It is to the latter that Elias Levita, who seems first to have made it known in print (1518), assigns its origin. It commonly appears as a problem relating to Turks and Christians, or to Jews and Christians, half of whom must be sacrificed to save a sinking ship.[2]

The next writer of note was Nozawa Teichō, who published his *Dōkai-shō* in 1664, and who followed the custom begun by Yoshida in the proposing of problems for solution. Nozawa solved all of Isomura's problems and proposed a hundred new ones. He also suggested the quadrature of the circle by cutting it into a number of segments and then summing these partial areas. He went so far as to suggest the same plan for the sphere, but in neither case does he carry his work to completion. It is of interest to see this approach to the calculus in Japan, contemporary with the like approach at this time in Europe. Muramatsu had approximated the volume of the

---

1 *De bello judaico,* III, 16. This was formerly attributed to Hegesippus of the second century A. D., but it is now thought to be by a later writer of uncertain date.

2 Common names are *Ludus Josephi, Josephsspiel, Sankt Peder's lek* (Swedish), and the *Josephus Problem.* The Japanese name was *Mameko-date*, the step-children problem. It was very common in early printed books on arithmetic, as in those of Cardan (1539), Ramus (1569), and Thierfelder (1587). The best Japanese commentary on the problem is Fujita Sadusuke's *Sandatsu Kaigi* (Commentary on *Sandatsu*), 1774.

sphere by means of the summation of cylinders formed on circles cut by parallel planes. He had taken 100 of these sections, and possibly more, and had taken some kind of average that led him to fix upon 524 as the volume of a sphere of radius 5. Nozawa apparently intends to go a step further and to take thinner laminae, thus approaching the method used by Cavalieri in his *Methodus indivisibilibus.*[1] It is possible, as we shall see later, that some hint of the methods of the West had already reached the Far East, or it is possible that, as seems so often the case, the world was merely showing that it was intellectually maturing at about the same rate in regions far remote one from the other.

Two years later, in 1666, the *annus mirabilis* of England, Satō Seikō[2] wrote his work entitled *Kongenki.* In this he attempted to solve the problems proposed by Isomura and Nozawa, and he set forth 150 new questions. Mention should also be made of his interest in magic circles. Since with him closes the attempts at the mensuration of the circle and sphere prior to the work of Seki, it is proper to give in tabular form the results up to this time.[3]

Author	Date	π	Area of Circle	Volume of sphere
Yoshida	1627	3.16	0.79	0.5625
Imamura	1639	3.162	0.7905	0.51
Yamada	1656	3.162	0.7905	0.4934
Shibamura	1657	3.162	0.7905	0.525
Isomura	1660	3.162	0.7905	0.51
Muramatsu	1663	3.14	0.785	0.524
Nozawa	1664	3.14	0.785	0.523
Satō	1666	3.14	0.785	0.519

[1] Written in 1629, but printed in 1635.

[2] Given incorrectly in FUKUDA's *Sampō Tamatebako* of 1879, and in ENDŌ, Book I, p. 73, as Satō Seioku.

[3] The table in substantially this form appears in HAYASHI's *History*, p. 37. See also HERZER, P., *loc. cit.*, p. 35 of the Kiel reprint of 1905; ENDŌ, I, p. 75.

Satō's *Kongenki* of 1666 is particularly noteworthy as being the first Japanese treatise in which the "Celestial element" method in algebra, as set forth in the *Suan-hsiao Chi-mêng*,[1] is successfully used. Some of the problems given by him require the solution of numerical equations of degree as high as the sixth, and it is here that Satō shows his advance over his predecessors. The numerical quadratic had been solved in Japan before his time, and even certain numerical cubics, but Satō was the first to carry this method of solution to equations of higher degree. In spite of the fact that he knew the principle, Satō showed little desire to carry it out, however, so that it was left to his successor to make more widely known the Chinese method and to show its great possibilities.

This successor was Sawaguchi Kazuyuki,[2] a pupil of Takahara Kisshu, and afterwards a pupil of the great Seki. In 1670 Sawaguchi wrote the *Kokon Sampō-ki*, the "Old and New Methods of Mathematics". The work consists of seven books, the first three of which contain the ordinary mathematical work of the time, and the next three a solution by means of equations of the problems proposed by Satō.[3] He also followed Nozawa in attempting to use a crude calculus (Fig. 27) somewhat like that known to Cavalieri. Sawaguchi was for a time a retainer of Lord Seki Bingo-no-Kami, but through some fault of his own he lost the position and the closing years of his life were spent in obscurity.[4]

Sawaguchi's solutions of Satō's problems are not given in full. The equations are stated, but these are followed by the answers only. An equation of the first degree is called a *kijo shiki*, "divisional expression", inasmuch as only division is needed in its solution, of course after the transposition and

[1] See p. 48.

[2] In later years he seems, according to the *Stories told by Araki*, to have changed his name to Gotō Kakubei, although other writers take the two to be distinct personages.

[3] It should also be mentioned that a similar use of equations is found in Sugiyama Teiji's work that appeared in the same year.

[4] *The Stories told by Araki.*

uniting of terms. Equations of higher degree are called *kaihō shiki*, "root-extracting expressions". As a rule only a single root of an equation is taken, although in a few problems this rule is not followed.[1] This idea of the plurality of roots is a

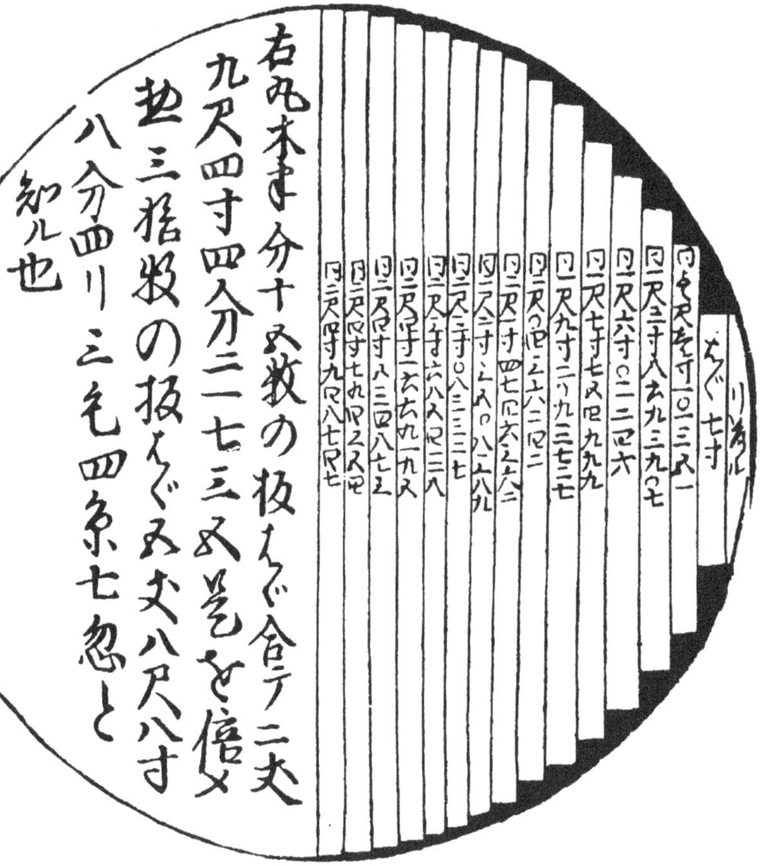

Fig. 27. Early steps in the calculus. From Sawaguchi Kazuyuki's *Kokon Sampō-ki* (1670).

noteworthy advance upon the work of the earlier Chinese writers, since the latter had recognized only one root to any equation. As is usual in such forward movements, however, Sawaguchi did not recognize the significance of the plural

[1] Satō had already recognised the plurality of roots in his *Kongenki*.

roots, calling problems which yielded them erroneous in their nature.

That Sawaguchi's methods may be understood as fully as the nature of his work allows, a few of his solutions of Satō's problems are set forth:

"There is a right triangle whose hypotenuse is 6, and the sum of whose area and the square root of one side is 7.2384. Required the other two sides". (No. 64.)

Sawaguchi gives the following direstions:

"Take the 'Celestial element' to be the first side. Square this and subtract the result from the square of the hypotenuse. The remainder is the square of the second side. Multiplying this by the square of the first side, we have 4 times the square of the area, which will be called A. Let 4 times the square of the first side be called B. Arrange the sum, square it, and multiply by 4. From the result subtract A and B. The square of the remainder is 4 times the product of A and B, and this we shall call X. Arrange A, multiply by B, take 4 times the product, and subtract the quantity from X, thus obtaining an equation of the 8th degree. This gives, evolved in the reverse method,[1] the first side." The result for the two sides are then given as 5.76, and 1.68.[2]

Satō's problem No. 16 is as follows: "There is a circle from within which a square is cut, the remaining portion having an area of 47.6255. If the diameter of the circle is 7 more than the square root of a side of the square, it is required to find the diameter of the circle and the side of the square."[3] Sawaguchi looks upon the problem as "deranged", since it has two solutions, viz., $d = 9$, $s = 4$, and $d = 7.8242133\ldots$ and $s = 0.67932764\ldots$. He therefore changes the quantities as given in

---

[1] That is, when the signs of the coefficients are changed in the course of the operation.

[2] Expressed in modern symbols, let $s =$ the sum, 7.2384, $h =$ the hypotenuse, and $x =$ the first side. Then, by his rule, $[4s^2 - (h^2 - x^2)x^2 - 4x^2]^2 - 16x^4(h^2 - x^2) = 0$.

[3] I. e., $\frac{1}{4}\pi d^2 - s^2 = 47.6255$, and $d - \sqrt{s} = 7$.

the problem, making the area 12.278, and the difference 4. He then considers the equation as before, viz., $\frac{1}{4}\pi d^2 - s^2 = 12.278$, and $d - \sqrt{s} = 4$. Then $d = 6$ and $s = 4$, taking $\frac{1}{4}\pi$ to be 0.7855.

Sawaguchi next considers a problem from the *Dōkai-shō* of Nozawa Teichō (1664), viz: "There is a rectangular piece of land 300 measures long and 132 measures wide. It is to be equally divided among 4 men as here shown, in such manner

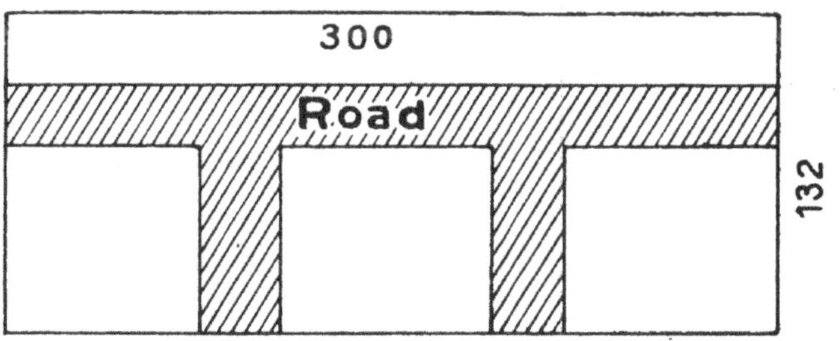

that three of the portions shall be squares. Required the dimensions of the parts."

Satō gives two solutions of this problem in his *Kongenki*, as follows:

1. Each of the square portions is 90 measures on a side; the fourth portion is 27 measures wide; and the roads are each 15 measures wide.

2. Each of the square portions is 60 measures on a side; the fourth portion is 12 measures wide; and the roads are each 60 measures wide.

This solution of Satō's leads Sawaguchi to dilate upon the subtle nature of mathematics that permits of more than one solution to a problem that is apparently simple.

Of the hundred and fifty problems in Satō's work Sawaguchi says that he leaves some sixteen unsolved because they relate to the circle. He announces, however, that it is his intention to consider problems of this nature orally with his pupils, and he gives without explanation the value of π as 3.142.

Two of the sixteen unsolved problems are as follows:—

"The area of a sector of a circle is 41.3112, the radius is 8.5, and the altitude of the segment cut off by a chord is 2. Required to find the chord." (No. 34.)

"From a segment of a circle a circle is cut out, leaving the remaining area 97.27632. The chord is 24, and the two parts

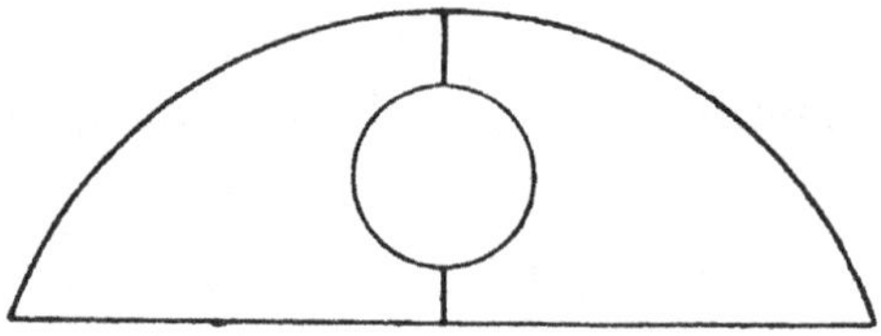

of the altitude, after the circle cuts out a portion as shown in the figure, are each 1.8. Required the diameter of the small circle."

The seventh and last book of Sawaguchi's work consists of fifteen new problems, all of which were solved four years later by Seki, who states that one of them leads to an equation of the 1458th degree. This equation was substantially solved twenty years later by Miyagi Seikō of Kyōto, in his work entitled *Wakan Sampō*.

CHAPTER VI.

Seki Kōwa.

In the third month according to the lunar calendar, in the year 1642 of our era, a son was born to Uchiyama Shichibei, a member of the *samurai* class living at Fujioka in the province of Kōzuke.[1] While still in his infancy this child, a younger son of his parents, was adopted into another noble family, that of Seki Gorozayemon, and hence there was given to him the name of Seki by which he is commonly known to the world. Seki Shinsuke Kōwa[2] was born in the same year[3] in which Galileo died, and at a time of great activity in the mathematical world both of the East and the West. And just as Newton, in considering the labors of such of his immediate predecessors as Kepler, Cavalieri, Descartes, Fermat, and Barrow, was able to say that he had stood upon the shoulders of giants, so Seki came at an auspicious time for a great mathematical advance in Japan, with the labors of Yoshida, Imamura, Isomura, Muramatsu, and Sawaguchi upon which to build. The coincidence of birth seems all the more significant because of the possible similarity of achievement, Newton having invented the calculus of fluxions in the West, while Seki possibly invented the *yenri* or "circle principle" in the East, each

---

1 Not far from Yedo, the Shogun's capital, the present Tōkyō.

2 Or Takakazu. On the life of Seki see MIKAMI, Y., *Seki and Shibukawa*, *Jahresbericht der Deutschen Mathematiker-Vereinigung*, Vol. XVII, p. 187; ENDŌ, Book II, p. 40; OZAWA, *Lineage of Mathematicians* (in Japanese); HAYASHI, *History*, part 1, p. 43, and the memorial volume (in Japanese) issued on the two-hundredth anniversary of Seki's death, 1908.

3 C. KAWAKITA, in an article in the *Honchō Sūgaku Kōenshū*, says that some believe Seki to have been born in 1637.

designed to accomplish much the same purpose, and each destined to material improvement in later generations. The *yenri* is not any too well known and it is somewhat difficult to judge of its comparative value, Japanese scholars themselves being undecided as to the relative merits of this form of the calculus and that given to the world by Newton and Leibnitz.[1]

Seki's great abilities showed themselves at an early age. The story goes that when he was only five he pointed out the errors of his elders in certain calculations which were being discussed in his presence, and that the people so marveled at his attainments that they gave him the title of divine child.[2]

Another story relates that when he was but nine years of age, Seki one time saw a servant studying the *Jinkō-ki* of Yoshida. And when the servant was perplexed over a certain problem, Seki volunteered to help him, and easily showed him the proper solution.[3] This second story varies with the narrator, Kamizawa Teikan[4] telling us that the servant first interested the youthful Seki in the arithmetic of the *Jinkō-ki*, and then taught him his first mathematics. Others[5] say that Seki learned mathematics from the great teacher Takahara Kisshu who, it will be remembered, had sat at the feet of Mōri as one of his *san-shi*, although this belief is not generally held. Most writers[6] agree that he was self-made and self-educated,

---

1 Thus ENDŌ feels that the *yenri* was quite equal to the calculus (*History*, Book III, p. 203). See also HAYASHI, *History*, part I, p. 44, and the *Honchō Sūgaku Kōenshū*, pp. 33—36. Opposed to this idea is Professor R. FUJISAWA of the University of Tōkyō who asserts that the *yenri* resembles the Chinese methods and is much inferior to the calculus. The question will be more fully considered in a later chapter.

2 KAMIZAWA TEIKAN (1710—1795), *Okinagusa*, Book VIII. KAMIZAWA lived at Kyōto. This title was also placed upon the monument to Seki erected in Tōkyō in 1794.

3 Kuichi Sanjin, in the *Sūgaku Hōchi*, No. 55.

4 *Okinagusa*, Book VIII.

5 See FUKUDA's *Sampō Tamatebako*, 1879; ENDŌ, Book II, p. 40; HAYASHI in the *Honchō Sūgaku Kōenshū*, 1908.

6 Fujita Sadasuke in the preface to his *Seiyō Sampō*, 1779; Ozawa Seiyō in his *Lineage of Mathematicians* (in Japanese), 1801; the anonymous manuscript entitled *Sanka Keizu*.

his works showing no apparent influence of other teachers, but on the contrary displaying an originality that may well have led him to instruct himself from his youth up.[1] Whatever may have been his early training Seki must have progressed very rapidly, for he early acquired a library of the standard Japanese and Chinese works on mathematics, and learned, apparently from the *Suan-hsiao Chi-mêng*,[2] the method of solving the numerical higher equation. And with this progress in learning came a popular appreciation that soon surrounded him with pupils and that gave to him the title of The Arithmetical Sage.[3] In due time he, as a descendent of the *samurai* class, served in public capacity, his office being that of examiner of accounts to the Lord of Kōshū, just as Newton became master of the mint under Queen Anne. When his lord became heir to the Shōgun, Seki became a Shogunate *samurai*, and in 1704 was given a position of honor as master of ceremonies in the Shōgun's household.[4] He died on the 24th day of the 10th month in the year 1708, at the age of sixty-six, leaving no descendents of his own blood.[5] He was buried in a Buddhist cemetery, the Jorinji, at Ushigome in Yedo (Tōkyō), where eighty years later his tomb was rebuilt, as the inscription tell us, by mathematicians of his school.

Several stories are told of Seki, some of which throw interesting sides lights upon his character.[6] One of these relates that he one time journeyed from Yedo to Kōfu, a city in Kōshū, or the Province of Kai, on a mission from his lord. Traveling in a palanquin he amused himself by noting the directions and

---

1 The fact that the long epitaph upon his tomb makes no mention of any teacher points to the same conclusion.

2 In the *Okinagusa* of Kamizawa this is given as the *Sangaku Gomō*, but in an anonymous manuscript entitled the *Sanwa Zuihitsu* the Chinese classic is specially given on the authority of one Saitō in his *Burin Inken Roku.*

3 In Japanese, *Sansei.* This title was also carved upon his tomb.

4 KAMIZAWA, *Okinagusa*, Book VIII; Kuichi Sanjin in the *Sūgaku Hōchi*, No. 55; ENDŌ, Book II, p. 40.

5 His heir was Shinshichi, or Shinshichirō, a nephew. ENDŌ, Book II, p. 81.

6 KAMIZAWA, *Okinagusa*, Book VIII.

distances, the objects along the way, the elevations and depressions, and all that characterized the topography of the region, jotting down the results upon paper as he went. From these notes he prepared a map of the region so minutely and carefully drawn that on his return to Yedo his master was greatly impressed with the powers of description of one who traveled like a *samurai* but observed like a geographer.

Another story relates how the Shōgun, who had been the Lord of Kōshū, once upon a time decided to distribute equal portions of a large piece of precious incense wood among the members of his family. But when the official who was to cut the wood attempted the division he found no way of meeting his lord's demand that the shares should be equal. He therefore appealed to his brother officials who with one accord, advised him that no one could determine the method of cutting the precious wood save only Seki. Much relieved, the official appealed to "The Arithmetical Sage" and not in vain.[1]

It is also told of Seki that a wonderful clock was sent from the Emperor of China as a present to the Shogun, so arranged that the figure of a man would strike the hours. And after some years a delicate spring became deranged, so that the figure would no longer strike the bell. Then were called in the most skilful artisans of the land, but none was able to repair the clock, until Seki heard of his master's trouble. Asking that he might take the clock to his own home, he soon restored it to the Shogun successfully repaired and again correctly striking the hours.

Such anecdotes have some value in showing the acumen and versatility of the man, and they explain why he should have been sought for a post of such responsibility as that of examiner of accounts.[2]

The name of Seki has long been associated with the *yenri*, a form of the calculus that was possibly invented by him, and

---

[1] The story is evidently based upon the problem of Yoshida already given on page 66.

[2] Kamizawa, *Okinagusa*, Book VIII.

that will be considered in Chapter VIII. It is with greater certainty that he is known for his *tenzan* method, an algebraic system that improved upon the method of the "Celestial element" inherited from the Chinese; for the *Yendan jutsu*, a scheme by which the treatment of equations and other branches of algebra is simpler than by the methods inherited from China and improved by such Japanese writers as Isomura and Sawaguchi, and for his work in determinants that antedated what has heretofore been considered the first discovery, namely the investigations of Leibnitz.

As to his works, it is said that he left hundreds of unpublished manuscripts,[1] but if this be true most of them are lost.[2] He also published the *Hatsubi Sampō* in 1674.[3] In this he solved the fifteen problems given in Sawaguchi's *Kokon Sampō-ki* of 1670, only the final equations being given.[4]

As to Seki's real power, and as to the justice of ranking him with his great contemporaries of the West, there is much doubt. He certainly improved the methods used in algebra, but we are not at all sure that his name is properly connected with the *yenri*.

For this reason, and because of his fame, it has been thought best to enter more fully into his work than into that of any of his predecessors, so that the reader may have before him the material for independent judgment.

First it is proposed to set forth a few of the problems that were set by Sawaguchi, with Seki's equations and with one of Takebe's solutions.

---

1 Endō, Book II, p. 41.

2 For further particulars see Endō, loc. cit., and the Seki memorial volume (*Seki-ryū Shichibusho*, or Seven Books on Mathematics of the Seki School) published in Tōkyō in 1908.

3 This is the work mentioned by Professor Hayashi as the *Hakki Sampō* of Mitaki and Mie (Miye).

4 In 1685 one of Seki's pupils, Takebe Kenkō, published a work entitled *Hatsubi Sampō Yendan Genkai*, or the "Full explanations of the *Hatsubi Sampō*," in which the problems are explained. He states that the blocks for printing the work were burned in 1680 and that he had attempted to make good their loss.

Sawaguchi's first problem is as follows: "In a circle three other circles are inscribed as here shown, the remaining area being 120 square units. The common diameter of the two smallest circles is 5 units less than the diameter of the one that is next in size. Required to compute the diameters of the various circles."

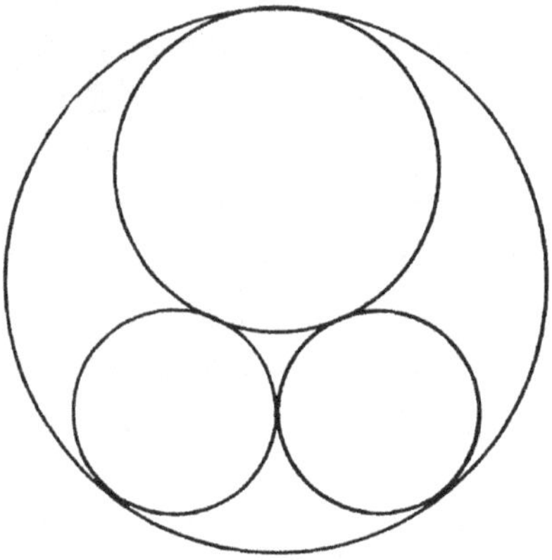

Seki solves the problem as follows: "Arrange the 'celestial element', taking it as the diameter of the smallest circles. Add to this the given quantity and the result is the diameter of the middle circle. Square this and call the result A.

"Take twice the square of the diameter of the smallest circles and add this to A, multiplying the sum by the moment of the circumference.[1] Call this product B.

"Multiply 4 times the remaining area by the moment of diameter.[2]

"This being added to B the result is the product of the square of the diameter of the largest circle multiplied by the moment of circumference. This is called C.[3]

---

1 By the "moment of the circumference" is meant the numerator of the fractional value of π. This is 22 in case π is taken as $\frac{22}{7}$.

2 "Moment of diameter" means the denominator of the fractional value of π. In the case of $\frac{22}{7}$, this is 7. That is, we have 7×120.

3 Thus far the solution is as follows: Let x = the diameter of the smallest circle, and y = the diameter of the largest circle. Then $x + 5$ is the diameter of the so-called "middle circle."

"Take the diameter of the smallest circle and multiply it by A and by the moment of the circumference. Call the result D.[1]

"From four times the diameter of the middle circle take the diameter of the smallest circle, and from C times this product take D. The square of the remainder is the product of the square of the sum of four times the diameter of the middle circle and twice the diameter of the smallest circle, the square of the diameter of the middle circle, the square of the moment of circumference, and the square of the diameter of the largest circle. Call this X.[2]

"The sum of four times the diameter of the middle circle and twice the diameter of the smallest circle being squared, multiply it by A and by C and by the moment of circumference.[3] This quantity being canceled with X we get an equation of the 6th degree.[4] Finding the root of this equation according to the reversed order[5] we have the diameter of the smallest circle.

"Reasoning from this value the diameters of the other circles are obtained."

Then $x^2 + 10x + 25 = A$,

$$22\,(3x^2 + 10x + 25) = B,$$

and $7 \cdot 4 \cdot 120 + B = C = 22y^2$, where $\pi = \frac{22}{7}$.

That the formula for C is correct is seen by substituting for 120 the difference in the areas as stated. We then have

$$7 \cdot 4 \cdot \frac{22}{7} \left\{\frac{y^2}{4} - \frac{(x+5)^2}{4} - \frac{2x^2}{4}\right\} + B = C,$$

$$\text{or } 22\,(y^2 - x^2 - 10x - 25 - 2x^2 + 3x^2 + 10x + 25) = C,$$

or $22y^2 = C$, which is, as stated in the rule, "the product of the square of the diameter of the largest circle multiplied by the moment of circumference."

[1] I. e., $22x\,(x^2 + 10x + 25) = D$.

[2] I. e., $\{C\,[4\,(x+5) - x] - D\}^2 = X$.

[3] I. e., $22 \cdot 22y^2\,(x+5)^2\,[4\,(x+5) + 2x]^2$. This is merely the second part of the preceding paragraph stated differently.

[4] I. e., $X = 22^2\,(3xy^2 + 5y^2 - x^2)^2$, and this quantity equals $22^2 y^2 (x+5)^2 (6x+20)^2$. Their difference is a sextic.

[5] As explained on page 53.

It may add to an appreciation or an understanding of the mathematics of this period if we add Takebe's analysis.

Let x be the diameter of the largest circle, y that of the middle circle, and z that of the smallest circles.[1]

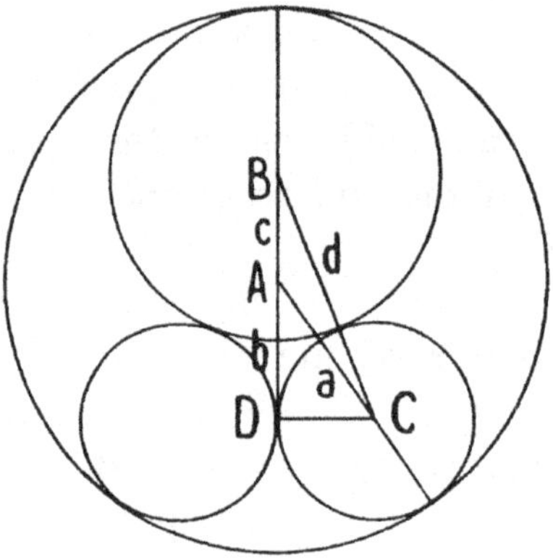

Then let $AC = a$, $AD = b$, $AB = c$, and $BC = d$, these being auxiliary unknowns at the present time.

Then

$$2a = -z + x,$$

and

$$4a^2 = z^2 - 2zx + x^2$$

or

$$4a^2 - z^2 = -2zx + x^2.$$

Therefore

$$4b^2 = -2zx + x^2. \qquad (1)$$

---

1 Takebe of course expresses these quantities in Chinese characters. The coefficients are represented by him in the usual *sangi* form, where |x, ∤y and ||xy stand respectively for x, $-y$, and $2xy$. This notation is called the *bōsho* or side-notation and is mentioned later in this work. Expressions containing an unknown are arranged vertically, and other polynomials are arranged horizontally. Thus for x, $-a+x$, $a^2-2ax+x^2$ we have

○	∤ a	| a^2
|	|	∦ a
		|

respectively, while for $a^2 + 2ab + b^2$ we have

|a^2 ||ab |b^2

with Chinese characters in place of these letters.

If we take y from x we have $-y + x$, which is $2c$. Squaring.

$$4c^2 = y^2 - 2yx + x^2. \tag{2}$$

To y add z and we have

$$2d = y + z.$$

Squaring,

$$4d^2 = y^2 + 2yz + z^2.$$

Subtracting z^2,[1] we have

$$4(b + c)^2 = y^2 + 2yz.$$

Subtract from this (1) and (2) and we have

$$b \times 8c = 2yz + (2z + 2y)x - 2x^2.$$

Dividing by 2,

$$b \times 4c = yz + (z + y)x - x^2.$$

Squaring,

$$b^2 \times 16c^2 = y^2z^2 + (2y^2z + 2yz^2)x + (y + z)^2x^2$$
$$- (2y + 2z)x^3 + x^4. \tag{3}$$

Multiplying (1) by (2) we also have

$$b^2 \times 16c^2 = -2y^2zx + (y^2 + 4yz)x^2 - (2y + 2z)x^3 + x^4,$$

which being canceled with the expression in (3) gives

$$y^2z^2 + (4y^2z + 2yz^2)x + (-4yz + z^2)x = 0,$$

from which, by canceling z,

$$y^2z + (4y^2 + 2yz)x + (-4y + z)x^2 = 0.$$

This may be written in the form

$$y^2z + (x^2z - 4x^2y) + (4y^2 + 2yz)x = 0.$$

Takebe has now eliminated his auxiliary unknowns, and he directs that the quantity in the first parenthesis be squared and canceled with the square of the rest of the expression,[2]

---

[1] And noting that $d^2 - (\frac{1}{2}) z^2 = (b + c)^2$.

[2] This amounts to equating $x^2z - 4x^2y$ to $-[y^2z + (4y^2 + 2yz)x]$, and then squaring and canceling out like terms.

and that the rest of the steps be followed as in Seki's solution. In this he expresses y and z in terms of x and given quantities and thus finds an equation of the sixth degree in x. Without attempting to carry out his suggestions, enough has been given to show his ingenuity in elimination.

The 12th problem proposed by Sawaguchi is as follows:

There is a triangle in which three lines, a, b, and c, are drawn as shown in the figure. It is given that

$$a = 4,\ b = 6,\ c = 1.447,$$

that the sum of the cubes of the greatest and smallest sides is 637, and that the sum of the cubes of the other side and of the greatest side is 855. Required to find the lengths of the sides.

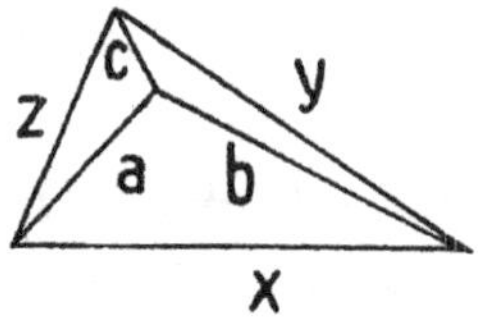

Seki solves this problem by the use of an equation of the 54th degree.

The 14th problem is of somewhat the same character. It is as follows:

There is a quadrilateral whose sides and diagonals are represented by u, v, w, x, y, and z, as shown in the figure.

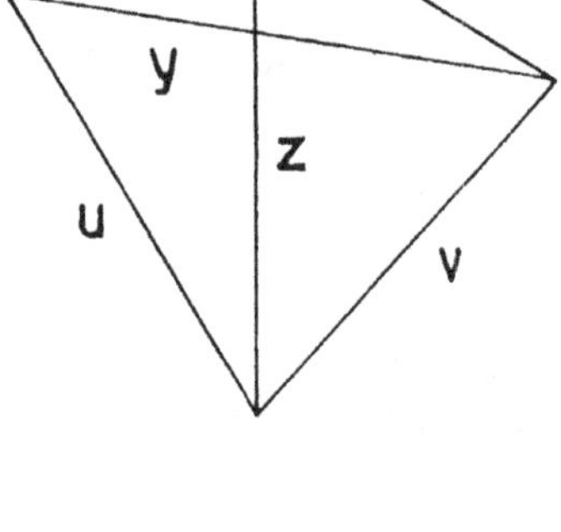

It is given that

$$z^3 - u^3 = 271$$
$$u^3 - v^3 = 217$$
$$v^3 - y^3 = 60.8$$
$$y^3 - w^3 = 326.2$$
$$w^3 - x^3 = 61.$$

Required to find the values of u, v, w, x, y, and z.[1]

Seki does not state the equation that is to be solved, but he says:

"To find z we have to solve by the reversed method an equation of the 1458th degree. But since the analysis is very complicated and cannot be stated in a simple manner we omit it, merely hinting at the solution.

"Take the 'celestial element' for z, from which the expressions of the cubes of u, v, w, x, and y may be derived.

"Then eliminate x^3, the analysis leading to an equation of the 18th degree.

"Next eliminate w^3, leading to an equation of the 54th degree.

"Next eliminate y^3, leading to an equation of the 162d degree.

"Next eliminate v^3, leading to an equation of the 486th degree.

"Now by eliminating u^3 two equal expressions result from which the final equation of the 1458th degree is obtained. Solving this equation by the reversed method we obtain the value of z. This method[2] of analysis leads us to the result step by step and may serve as an example of the method of attacking difficult problems."

Seki's explanation is, as he states, very obscure. Undoubtedly he explained the work orally to his pupils, with the *sangi* at hand. As the matter stands in his statement it would appear that he had five equations with six unknowns and that he had

---

[1] This is exactly as in the original, except that symbols replace the words. With merely these equations it is indeterminate. Takebe adds another equation, $z^2 + u^2 - x^2 = z \cdot 2s$, where s is the projection of u upon z.

[2] Essentially the method of constructing the equation.

not made use of the geometric relations involved, so that we are left to conjecture what particular equations he may have employed.

Although the explanations given by Seki, as shown in the few examples quoted, are manifestly incomplete and obscure, they are nevertheless noteworthy as marking a step in mathematical analysis. His predecessors had been content to state mere rules for attaining their results, as were also many of the early European algebraists. Leonardo of Pisa, for example, solves a numerical cubic equation to a remarkable degree of approximation, but we have not the slightest idea of his method. Even in the sixteenth century the Italian and German algebraists were content to use the Latin expression "Fac ita".[1] Seki, however, paid special attention to the analysis of his problems, and to this his great success as a teacher was largely due. His method of procedure was known as the *yendan jutsu*, *yendan* meaning explanation or expositon, and *jutsu* meaning process,[2] a method in which the explanation was carried along with the manipulating of the *sangi* in the "Celestial Element" calculation of the Chinese. When a problem arises in which two or more unknowns appear there are, in general, two or more expressions involving these unknowns. These expressions Seki was wont to write upon paper, and then to simplify the relations between them until he reached an equation that was as elementary in form as possible. This was in opposition to the earlier plan of stating the equation at once without any intimation of the method by which it was derived. Moreover it led the pupil to consider at every step the process of simplifying the work, thus reducing as far as possible the degree of the equation which was finally to be solved.[3] Seki's pupil, Takebe, speaks enthusiastically of his master's clearness of analysis, in these

---

[1] In early German, *thu ihm also.*

[2] We might translate the expression by the single word analysis.

[3] ENDŌ calls attention to the fact that the *yendan jutsu* may be looked upon as the repeated application of the *tengen jutsu* mentioned on p. 48. See his *Biography of Seki* (in Japanese) in the *Tōyō Gaku-gei Zasshi*, vol. 14, p. 313.

words:[1] "In fact this *yendan* is a process that was never set forth in China with the same clearness as in Japan. It is one of the brilliant products of my master's school and it must be agreed that it surpasses all other mathematical achievements, ancient or modern."

These words seem to be those of an enthusiastic disciple rather than a simple chronicler of fact, since from the evidence that is before us the *yendan* was merely a common-sense form of analysis such as any mathematician or teacher might employ, although we must admit that his predecessors had not made any use of it.

Takebe is not content, however, to let Seki's fame as a teacher rest here, and so he hints at another and rather esoteric theory, as one of the initiates of the Pythagorean brotherhood might have given mysterious reference to some carefully concealed principle of the great master.

"Although", he says, "there is yet another divine method that is more far-reaching, still I shall not attempt to explain it, for fear that one whose knowledge is so limited as mine would tend to misrepresent its significance,"—a tribute, probably, to the *tenzan* method, Seki's improvement upon that of the "Celestial Element".[2] Takebe's reticence in speaking of it may merely have reflected the modesty of Seki himself, for of this modesty we are well assured by divers writers. To boast of such an invention would have been entirely foreign to the *samurai* spirit of Seki and to the exalted principles of *Bushidō*. On the other hand, this custom of secrecy had existed everywhere before Seki's time, as witness the attitude of Tartaglia and Cardan, and even of a man like Galileo. In Japan, Mōri is said to have kept a secret book that was revealed only to his most deserving pupils,[3] and Isomura also had one, his

---

1 TAKEBE, *Hatsubi Sampō Yendan Genkai*, 1685, preface.

2 *Tenzan* has a broader meaning that may here be understood. It includes practically all of Japanese mathematics except possibly *yenri*. In a restricted sense it is written mathematics, but it sometimes includes the "Celestial Element" method.

3 See the *Sanwa Zuihitsu*.

book treating of the calculations relating to a circle and an arc.[1] Seki was so impressed with his discovery that he revealed it to his most promising followers only upon their swearing, with their own blood, never to make it public. And so, for more than half a century after Seki's death the secret remained, not becoming known to the world until Arima Raidō, feudal lord[2] of Kurume, in the island of Kyūshū, revealed it in his *Shūki Sampō*[3] in 1769.

This method was called by Seki the *kigen seihō*, meaning a method for revealing the true and buried origin of things. The term suggests the title of the papyrus of Ahmes, written in Egypt more than three thousand years earlier, "The science of dark things." It would be interesting to know the origin and history of this name for algebra or certain algebraic processes, since it is found in various parts of the world and in various ages. The *tenzan* method being the one to which Takebe seems to have referred in his work of 1685, we are quite certain that it was invented some time before this date.[4] It is first called by this name by Matsunaga Ryōhitsu. It is related that Lord Naitō of Nobeoka, in Kyūshū, himself no mean mathematician, was the one who caused the adoption of the name, requiring Matsunaga, a pupil of Araki who was a direct disciple of Seki, to write the *Hōrō-Yosan* in which it appears.[5]

The word *tenzan* consists of two Chinese ideograms, *ten* meaning to restore, and *zan* meaning to strike off. It would be most interesting if we could know the relation (if any) between this term and the name given by Mohammed ibn Musa al-Khowarazmi (c. 830) to his algebra,—*al-jebr w'al-muqabala*, which words mean substantially the same thing,—

---

1 Ibid.

2 Daimyō.

3 It was in this book that the value of π to fifty decimal places was first printed in Japan, an approximation already reached by Matsunaga.

4 Endō, in the *Tōyō Gaku-gei Zasshi*, vol. 14, p. 314.

5 Ozawa's *Lineage of Mathematicians* (Japanese), 1801. The *Hōrō-Yosan* is a manuscript without date.

restoration and reduction.[1] Does this resemblance tell of the passing of the mystery of "the science of dark things" from one school to another in the perpetual interchange of thought in the world's great republic of scholars, or are these resemblances that are continually met in the history of mathematics mere coincidences? This *tenzan* method may, however, justly be called a purely Japanese product, the product of Seki's brain, and quite unrelated to any Chinese treatment.[2]

We shall now speak of the notation employed in this method. This notation is the *bōsho shiki* already mentioned. In earlier times it had been the habit of Japanese mathematicians to represent numbers by the *sangi* method described in Chapter IV and known as the *chū-shiki*.[3] Seki amplifies this by writing the numerals at the side of a vertical line, the significance of which will be explained in a moment. Since these numerals were written at the side of a line this method of writing them is known as *bōsho shiki* or "side notation". In our explanation we necessarly use Latin letters and Hindu-Arabic forms instead of the Chinese ideograms, but otherwise the representations are substantially correct. Seki writes $\frac{2}{3}$, $\frac{1}{n}$, and $\frac{abc}{mn}$ as follows: 3|2, *n*| or *n*|1, *mn*|*abc*, the numerators being placed on the right and the denominators on the left. Sometimes the vertical line is replaced by *sangi* coefficients, as in the case of ||||*ab*, *r*||π, 27≡|||||*abc*, for 4 ab, $\frac{2\pi}{r}$, and $\frac{35abc}{27}$.

Powers of quantities are represented thus:

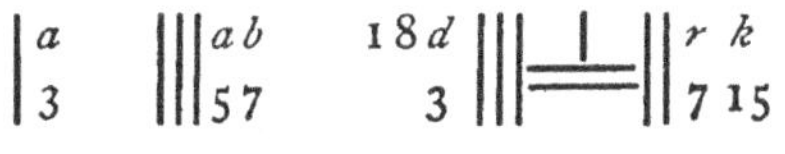

for a^4, $3a^6b^8$, $\frac{372r^8k^{16}}{18d^4}$. It will be seen that the exponent in each case is one less than that used in occidental mathematics.

---

[1] The varied fortunes of the name for algebra, in Europe, is interesting. Thus we have such titles as *algiebr, algobra, mukabel, almucable, arte maggiore, ars magna, coss, cossic art,* and so on.

[2] Endō, Book II, p. 8.

[3] *Sangi* notation.

The reason is that in the *wasan* as in Chinese mathematics the *n*th power of a quantity is called the "(n—1) times self-multiplied". That is, the native oriental exponent shows not the number of factors but the number of times a quantity is multiplied by itself. The fractional exponent was not used in the native algebra of Japan.

The "side notation" was also used in other ways. Thus $a + b$ might be indicated in either of the ways here shown.

$$\begin{array}{l} |a \\ |b \end{array} \quad \text{or} \quad |a \ |b$$

To indicate subtraction an oblique cancelation line was used. Thus *b*—*a* was indicated in these four ways:

$$\begin{array}{l} \nmid a \\ |b \end{array} \quad \begin{array}{l} |b \\ \nmid a \end{array} \quad \nmid a|b \quad |b\times a$$

It will be noticed that this *tenzan* notation was employed in Seki's *yendan* method. Indeed the *tenzan* may be considered as the notation, while the *yendan* refers to the method of analysis. It is difficult to justify the extravagant praise of the disciples of Seki with respect to either of these phases of his work. He must have been very clear in his own analysis with his pupils, and this gave them a higher appreciation of the *yendan* than anything that has come down to us would warrant. And as for the notation, while this is an improvement upon that of the Chinese, the improvement does not seem to have been so great as to warrant the praise which it has provoked. It was applied to the entire range of Japanese mathematics except the *yenri* or circle principle,[1] but we know that the Chinese notation would have been quite sufficient for the work to be accomplished. In its application to factoring, the finding of highest common factor and the lowest common multiple, the summation of infinite series and of power series of the type $1^n + 2^n + 3^n + \ldots$, the *shōsa-hō* or method of differences, the theory of numbers, the *tetsu-jutsu* or expansion in series of the root of a quadratic equation, the calculation relating to

---

[1] See ARIMA's *Shūki Sampō,* 1769; ENDŌ, Book II, pp. 4, 5, and in the *Tōyō Gaku-gei Zasshi,* vol. 14, pp. 362—364.

regular polygons, and the study of maxima and minima, the *tenzan* notation seems to have served its purposes fairly well, better indeed than any notation known in Japan up to that time. How much of this application to the various branches of algebra was due to Seki and how much to his disciples, we shall never know. The old Pythagorean idea of *ipse dixit* seems to have prevailed in Seki's school, and the master may often have received credit for what the pupil did.

Thus far, indeed, we have not found much in the way of discovery to justify the high standing of Seki. It is therefore well to consider some of the more serious contributions attributed to him. For this purpose we shall go to a work published by Ōtaka Yūshō in 1712, although compiled before 1709, that is, soon after Seki's death. Ōtaka was a pupil of Araki Sonyei, who had learned from Seki himself, and the book claims to be a posthumous publication of the works of this master, edited by Ōtaka under Araki's guidance. Although this work, known as the *Katsuyō Sampō*,[1] does not contain the *tenzan* system, it gives a good idea of some of Seki's other work, and on this account the publication was a subject of deep regret to the brotherhood of his followers. Tradition says that it was owing to the protests of these followers that no further publication of Seki's works was undertaken at a time when an abundance of material was at hand.

One of the subjects treated in the *Katsuyō Sampō* is the *shōsa-hō* or *shōsa* method, a theory that seems to have arisen from the study of problems like the summation of $1^n + 2^n + 3^n + \ldots$ Suppose, for example, we have such a function as

$$P = a_1 x + a_2 x^2 + \ldots + a_n x^n,$$

where the coefficients are as yet undetermined. Then if a sufficient number of values P_i are known for various values of x, the various values a_i can be determined, and this is one of the problems of the *shōsa-hō*. Professor Hayashi speaks of the method in general as that of finite differences, and this certainly is one of its distinguishing features.

---

1 "A summary of arithmetical rules."

This *shōsa-hō* in its general form is not an invention of Seki's. It appears to be of Chinese origin, perhaps invented by Kuo Shou-ching, a celebrated astronomer of the court of the Mogul Empire of the 13th and 14th centuries, and possibly even of earlier origin. There are three special forms, however: (1) the *ruisai shōsa* of which an illustration has just been given; (2) the *hōtei shōsa*, and (3) the *konton shōsa*, these latter two being first described in the *Shūki Sampō* of 1769. Seki's contribution was, therefore, a worthy generalization of an older Chinese device, and the application of this improvement to new problems.

The *shōsa-hō* was doubtless employed by Ōtaka in his *Katsuyō Sampō* (1712), in which there appears a table that expresses the formulas for the power series

$$S_r = 1^r + 2^r + 3^r + \dots + n^r,$$

for $r = 1, 2, 3, \dots N$. Such power series were called by the name *hōda*, and some of the results of their summation are as follows:

$$S_1 = \frac{1}{2}(n^2 + n),$$

$$S_2 = \frac{1}{6}(2n^3 + 3n^2 + n),$$

$$S_3 = \frac{1}{4}(n^4 + 2n^3 + n^2),$$

$$S_4 = \frac{1}{30}(6n^5 + 15n^4 + 10n^3 - n),$$

$$S_5 = \frac{1}{12}(2n^6 + 6n^5 + 5n^4 - n^2),$$

and so on to

$$S_{11} = \frac{1}{24}(2n^{12} + 12n^{11} + 22n^{10} - 33n^8 + 44n^6 - 33n^4 + 10n^2).$$

In Book III of this same work, the *Katsuyō Sampō*, there is his *Kakuhō narabini Yendan-Zu*, a treatment of the subject of regular polygons, namely of those of sides numbering 3, 4, ... 20. To illustrate some of the results we shall consider the case of the apothem of a regular polygon of thirteen sides.

Using the annexed figure, as given in the *Katsuyō Sampō* (see Fig. 28 for the original), and letting the side of the

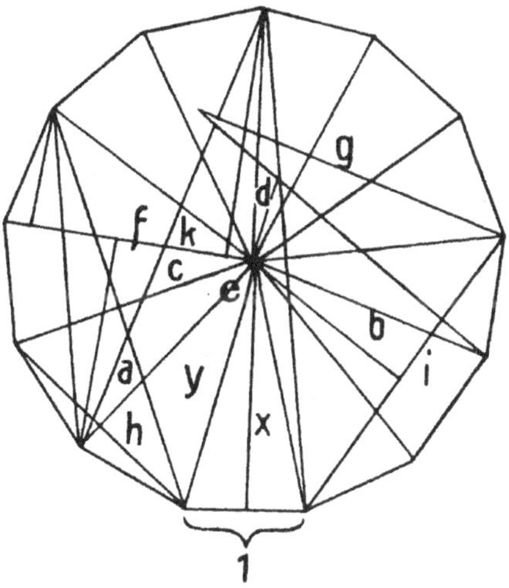

polygon be unity, the apothem x, and the radius y, we have

$$1^2 + 4x^2 = 4y^2.$$

Now

$$(1 + 4x^2)^3 = 1 + 12x^2 + 48x^4 + 64x^6 = 4096xabcde,$$

a statement made without any explanation. Ōtaka now proceeds by a series of unproved statements to develop two equations, viz.,

$$-1 + 312x^2 - 114{,}400x^4 + 109{,}824x^6 - 329{,}472x^8 + 292{,}864x^{10} - 53{,}248x^{12} = 0,$$

from which we are to find x, the apothem, and

$$-1 + 13y^2 - 65y^4 + 156y^6 - 182y^8 + 91y^{10} - 13y^{12} = 0,$$

from which we are to find y, the radius.

The treatment of the circle is given in Book IV of the *Katsuyō Sampō* and is similar to that attempted by Muramatsu in his *Sanso* of 1663. A circle of unit diameter is taken, a square is inscribed, and the sides of the inscribed regular polygon are continually doubled until a polygon of 2^{17} sides is reached.

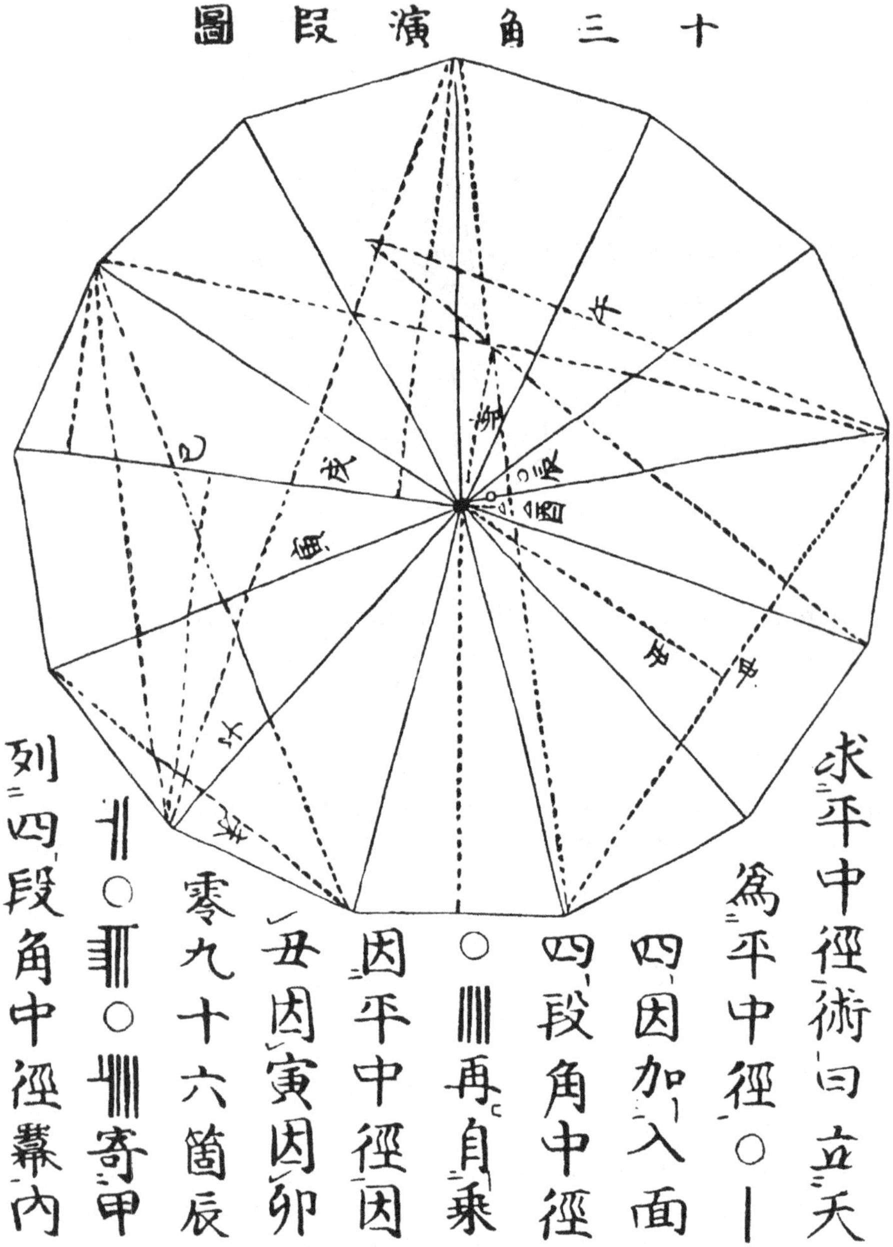

Fig. 28. From Ōtaka's *Katsuyō Sampō* (1712).

The treatment thus far is not at all original, but the work is carried farther than in Muramatsu's treatise and it represents about the same state of mathematical progress that was found in Europe some fifty years earlier than Muramatsu, or about a century before the death of Seki. Two new features, however, appear. Of these the first is that if the perimeters of the last three polygons are

$$a = 3.\ 14159\ 26487\ 76985\ 6708 -$$

$$b = 3.\ 14159\ 26523\ 86591\ 3571 +$$

$$c = 3.\ 14159\ 26532\ 88992\ 7759 -$$

then

$$\pi = b + \frac{(b-a)\,(c-b)}{(b-a)-(c-b)}$$

$$= 3.\ 14159265359 -,$$

which reminds us of some of the incorrect assumptions of the Antiphon-Bryson period, and of the close of the sixteenth century in Europe.

The second feature is, however, the interesting one. Starting with the fraction $\frac{3}{1}$, if we increase the denominator successively by unity, and then increase the numerator successively by 4 or by 3 according as the previous fraction is less or greater than the known decimal value of π, we shall obtain a series of values as follows:

(1)	$\frac{3}{1} = 3,$	"Old value,"	less than π
(2)	$\frac{7}{2} = 3.5,$		greater than π
(3)	$\frac{10}{3} = 3.33\ldots,$		" " "
(4)	$\frac{13}{4} = 3.25,$		" " "
(5)	$\frac{16}{5} = 3.2,$		" " "
(6)	$\frac{19}{6} = 3.166\ldots,$		" " "
(7)	$\frac{22}{7} = 3.142857\ldots,$	"Exact value,"	" " "

(8) $\frac{25}{8} = 3.125$,	"Chih's value,"		less than π
(20) $\frac{63}{20} = 3.15$,	"T'ung Ling's value,"		greater than π
(25) $\frac{79}{25} = 3.16$,	"Old Japanese value,"		" " "
(45) $\frac{142}{45} = 3.155\ldots$,	"Liu Chi's value,"		" " "
(50) $\frac{157}{50} = 3.14$,	"Hui's (Liu Hui's) value,"		less than π
(113) $\frac{355}{113} = 3.14159292\ldots$,			greater than π

The names above quoted are given by Ōtaka, and are probably those used by Seki. The last value, $\frac{355}{113}$, is not assigned a name, which seems to show that Seki was not aware of Tsu Ch'ung-chih's measurement of the circle as set forth in his *Chui-shu*, and recorded in Wei Chih's *Sui-Shu.*[1] The value itself first appears in printed form in Japan in the works of Ikeda Shōi (1672), Matsuda Seisoku (1680) and Takebe Kenkō (1683).

The problem of computing the length of a circular arc also appears in the *Katsuyō Sampō*, the formula being given as

$$\begin{aligned}1276900\,(d-h)^5 a^2 = {} & 5107600\,d^6 h - 23835413\,d^5 h^2 \\ & + 43470240\,d^4 h^3 - 37997429\,d^3 h^4 \\ & + 15047062\,d^2 h^5 - 1501025\,d h^6 \\ & - 281290\,h^7,\end{aligned}$$

where $d =$ diameter, $h =$ height of segment, and $a =$ length of arc. In the special case where $d = 10$ and $h = 2$ this reduces to

$$41841459200\,a^2 = 3597849073280.$$

The method[2] of deriving this formula seems to have been purely inductive, the result of repeated measurements, since the explanation is so obscure as to be entirely unintelligible.

---

1 "Records of the Sui Dynasty." This fact was known, however, to Takebe, who mentions it in his *Fukyū Tetsujutsu* of 1722. It is also given in Matsunaga's *Hōyen Sankyō* of 1739. See also p. 14, above. The original *Chui-shu* of Tsu Ch'ung-chih has been lost.

2 Perhaps relates to the *shōsa* method in a modified form.

The volume of the sphere is computed in the *Katsuyō Sampō* (and also in Seki's *Ritsuyen-ritsu-Kai*) in an ingenious manner. The sphere is cut into 50, 100, and 200 segments of equal altitude, the diameter being taken as 10. From this Ōtaka obtains in some way the three parameters 666.4, 666.6, 666.65, each of which he multiplies by $\frac{\pi}{4}$ to obtain the three volumes. Calling the parameters a, b, and c, he now takes a mean in this manner:

$$b + \frac{(b-a)\ (c-b)}{(b-a) - (c-b)} = 666\frac{2}{3},$$

as in the case of the circle. Multiplying by

$$\frac{\pi}{4} = \frac{355}{4 \times 113}, \text{ we have}$$

$$666\frac{2}{3} \times \frac{355}{4 \times 113} = 523\frac{203}{339} = \frac{355}{678} \times 1000$$

for the required volume. This amounts to taking $\frac{355}{678}$ for $\frac{\pi}{6}$, which means that the formula $v = \frac{4}{3}\pi r^3$ is correctly used.

One of Seki's favorite studies was the theory of equations, a subject treated in his works on the *Kaihō Hompen*,[1] the *Byōdai Meichi*,[2] the *Daijutsu Bengi*[3], the *Kaihō Sanshiki*[4] and the *Kaihō Hengi-jutsu*.[5] In the first of these works he classifies equations into four kinds, the *jenshō shiki* (perfect equations), *henshō shiki* (varied equations), *kōshō shiki* (mixed equations), and the *mushō shiki* (rootless equations), a system not unlike those found in the works of the Persian and Arabian writers, the classification according to degree being relatively modern even in Europe. By a perfect equation he means one that has only a single root, positive or negative. A varied equation is one in which several roots occur, but all of the same sign. A mixed equation is one in which several roots

---

1 "Various topics about equations."

2 Literally, "On making pathological problems perfect."

3 Literally, "Discussion on the data of problems."

4 Literally, "Considerations on the solution of equations."

5 Literally, "On new methods for the solution of equations."

occur, but not all of the same sign. A rootless equation is one having neither a positive nor a negative root, restricted as Seki was aware to equations of even degree.[1]

In the *Kaihō Hompen*[2] Seki treats of positive and negative roots, and sets forth a method called the *tekizin-hō*[3] represented by the following table:

0 degree	1	1	1	1	1	1	1
1st. „		1	2	3	4	5	6
2d. „			1	3	6	10	15
3d. „				1	4	10	20
4th. „					1	5	15
5th. „						1	6
6th. „							1

The method of deriving this table, analogous to that for the Pascal Triangle, is evident. Indeed, the vertical columns are simply the horizontal ones of the usual triangular array. Seki does not tell how the numbers are obtained, and no explanation seems to have been given by any Japanese until Wada Nei gave one in the first half of the nineteenth century.[4] Such an array is rather obvious and was known long before Pascal or even Apianus (1527) published it.[5] Seki might have used it, as others in the West had done, for binomial coefficients, but it was not meant by him for this purpose.

In his *Byōdai Meichi* Seki calls attention to the fact that

---

1 I. e., in general. Of course we have also $x = \sqrt{-2}$, $x = \pi i$, etc., as well as $x^3 = \sqrt{-2}$, etc., although Seki makes no mention of such forms, having apparently no conception of the imaginary root.

2 The Kaihō-Houpen of Hayashi's *History*, part I, p. 52.

3 Literally, "Vanishing method," relating to maxima and minima.

4 In connection with his theory of maxima and minima.

5 SMITH, D. E., *Rara Arithmetica*, Boston, 1908, p. 155.

the mensuration of the circle or of any regular polygon requires but a single given quantity; that of a rectangle or pyramid, two given quantities; and that of a trapezoid, three. He then designates as *tendai* (insufficient problems) those problems in which there are not enough data for a solution, while those having too many data are designated as *handai* (excessive problems). He also states that in certain problems, although the data are correct as to number, no perfect answer is to be expected, and these problems he calls *kyodai* (imaginary). They arise, he says, in three cases: (1) where there is no root, (2) where all roots are negative, and (3) where the roots of the equation do not satisfy the conditions of the original problem. To illustrate the latter case he uses a simple problem involving the elementary principle of geometric continuity. He proposes to find the greater base of a trapezoid of altitude 9, the difference between the bases being 4, and the smaller base being 10 less than the altitude. The problem is trivial, the smaller base being 9—10 or —1, and the greater being 4—1 or 3. The smaller base, —1, does not appear to Seki to satisfy a geometric problem, so he proceeds with considerable circumlocution to explain what is perfectly obvious, that the trapezoid is a cross quadrilateral. The question of possible roots of an equation is discussed at some length but in a very elementary manner.

Problems leading to equations with two or more roots, or with negative roots, or with positive roots that do not satisfy the conditions of the problems, are called by Seki *hendai* or pathological problems, and were intended to be transformed into the ordinary determinate cases by a change in the wording.

In his solution of numerical equations Seki not only used the "celestial element" plan by which the Chinese had anticipated Horner's Method as early as 1247, but he effected at least one improvement on the Chinese plan,[1] unconsciously following a line laid down by Newton.

---

[1] This is seen in two manuscript works entitled *Kaihō Sanshiki* and *Kaihō Hengi-jutsu*.

For example, in the equation

$$11 + 8x + x^2 = 0,$$

the "celestial element" method gives the first two figures of one root as —1.7. Proceeding as usual in Horner's Method we have an equation of the form

$$0.29 + 4.6x + x^2 = 0.$$

Seki now takes $\frac{0.29}{4.6} = 0.063$, but unlike his predecessors he treats this as negative since the two coefficients are positive, and proceeds as before, his next equation being of the form

$$0.004169 + 4.474x + x^2 = 0.$$

Repeating the process we have for the continuation of the root —0.0009318. Continuing the same process Seki obtains for the root —1.76393202250020.

One of Seki's Seven Books[1] is devoted to magic squares and circles, a subject to which he may have been led by his study (in 1661) of a Chinese work by Yang Hui. He considers separately the magic squares with an odd number and an even number of cells, and with him begins the first scientific, general treatment of the subject in Japan. He begins by putting into obscure verse his rule for arranging a square of 3^2 cells. It would have been impossible to make out the meaning had Seki not given the square in a subsequent part of his manu-

4	9	2
3	5	7
8	1	6

script. As here shown the square is the common one that was well known long before Seki's time. Upon his method

[1] The *Hōjin Yensan*, (*Hōjin Ensan*) revised in manuscript in 1683. Araki gave to these the name of "Seven Books" (*Shichibusho*), and these he handed down to his disciples.

for a square of 3^2 cells he bases his general rule for one of $(2n+1)^2$ cells, and this is substantially as follows:

Begin with the cell next to the left of the upper right-hand corner and number to the right and down the right-hand

12	11	10	5	4	1	2
47						3
44						6
43						7
42						8
41						9
48	39	40	45	46	49	38

column until n is reached. In the annexed figure we have a square of

$$(2n+1)^2 = (2.3+1)^2 = 7^2 \text{ cells.}$$

We therefore number until 3 is reached. Then go to the left, from the cell to the left of 1, until $2n-1$ (in this case $2.3-1=5$) is reached. Then continue down the right side to the cell preceding the lower right-hand one, giving 6, 7, 8, 9. Then continue along the top row until the upper left-hand corner is reached, giving 10, 11, 12. This leaves the left-hand column to be completed, and the lower row to be filled. This is done by filling all except the corner cells by the complements to $(2n+1)^2+1$ of the respective numbers on the opposite side, — in this case the complements to the number 50. Thus, $50-3=47$, $50-6=44$, and so on. The corner cells are complements to 50 of the opposite corners.

The next step is to take n figures to the left of the upper right-hand corner and interchange them with the corresponding ones in the lower row, and similarly for the n figures above the lower right hand corner. The square then appears as here shown.

12	11	10	45	46	49	2
47						3
44						6
7						43
8						42
9						41
48	39	40	5	4	1	38

To fill the inner cells Seki follows a similar rule, except that the numbers now begin with 13. Without entering upon the exact details it will be easy for the reader to trace the plan by studying the result as here shown. The innermost square of 3^2 cells is filled by the method first given.

12	11	10	45	46	49	2
47	20	19	35	37	14	3
44	34	24	29	22	16	6
7	17	23	25	27	33	43
8	18	28	21	26	32	42
9	36	31	15	13	30	41
48	39	40	5	4	1	38

The even-celled squares have always proved more troublesome than the odd-celled ones. Seki first gives a rule for a square of 4^2 cells, with the result as here shown. He then divides these squares into those that are simply even and those that are doubly even.[1]

4	9	5	16
14	7	11	2
15	6	10	3
1	12	8	13

For the simply even squares above 4^2, Seki begins with the third cell to the left of the upper right-hand corner, proceding thence to the left, as shown in the figure. Then he goes back to the upper right-hand cell (for 5, in the case here shown) and proceeds down the right-hand column to the third cell from the bottom. He then fills the vacant cell at the top

4	3	2	1	9	5
31					6
30					7
29					8
27					10
32	34	35	36	28	33

(in this case with 9), and puts the next number (10) in the next cell in the right-hand column. The remaining cells in the left-hand column and the lower row are complements of the corresponding numbers with respect to $4(n+1)^2+1$, there being $2(n+1)$ elements on a side, as in the case of an odd-celled square. The interchange of elements is now made in a manner somewhat like that of the odd-celled square,

[1] $[2(n+1)]^2$, and $[2(2n)]^2$.

the result being here shown for the case of a square of 6^2 cells. The rest of the process is as in the odd-celled case.

4	3	35	36	28	5
6					31
30					7
8					29
10					27
32	34	2	1	9	33

For the doubly even magic square the first step of Seki's method will be sufficiently understood by reference to the following figure, in which the number is 8^2. The inner squares are filled in order until the one of 4^2 cells is reached, when that is filled in the manner first shown.

6	5	4	3	2	1	8	7
56							9
55							10
54							11
53							12
52							13
51							14
58	60	61	62	63	64	57	59

Seki simplified the treatment of magic circles, giving in substance the following rule:

Let the number of diameters be n. Begin with 1 at the center and write the numbers in order on any radius, and so on along the next $n-1$. Then take the radius opposite the last one and set the numbers down in order, beginning at the outside, and so on along the rest of the radii. In Fig. 29 the sum on any circle is 140, and for readers who have not become familiar with the Chinese numerals the following diagram, although arranged for only thirty three numbers, will be of service:

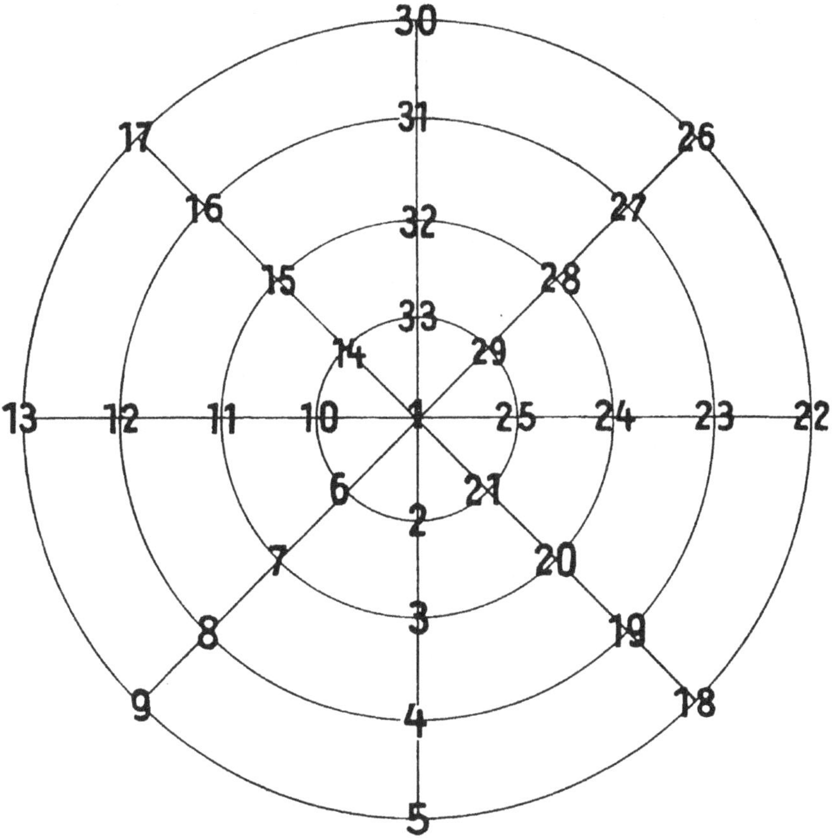

In another of Seki's manuscripts[1] there appears the Josephus problem already mentioned in connection with Muramatsu.

Mention should be made of Seki's work on the mensuration of solids, which appears in two of his manuscripts.[2] He begins

[1] *Sandatsu Kempu* (*Kenpu*).

[2] The *Kyūseki* (Calculation of Areas and Volumes) and the *Kyūketsu Hengyō Sō* (An incomplete treatise on the volume of a sphere).

by considering the volume of a ring[1] generated by the revolution of a segment of a circle about a diameter parallel to the chord of the segment. He states that the volume is equal to

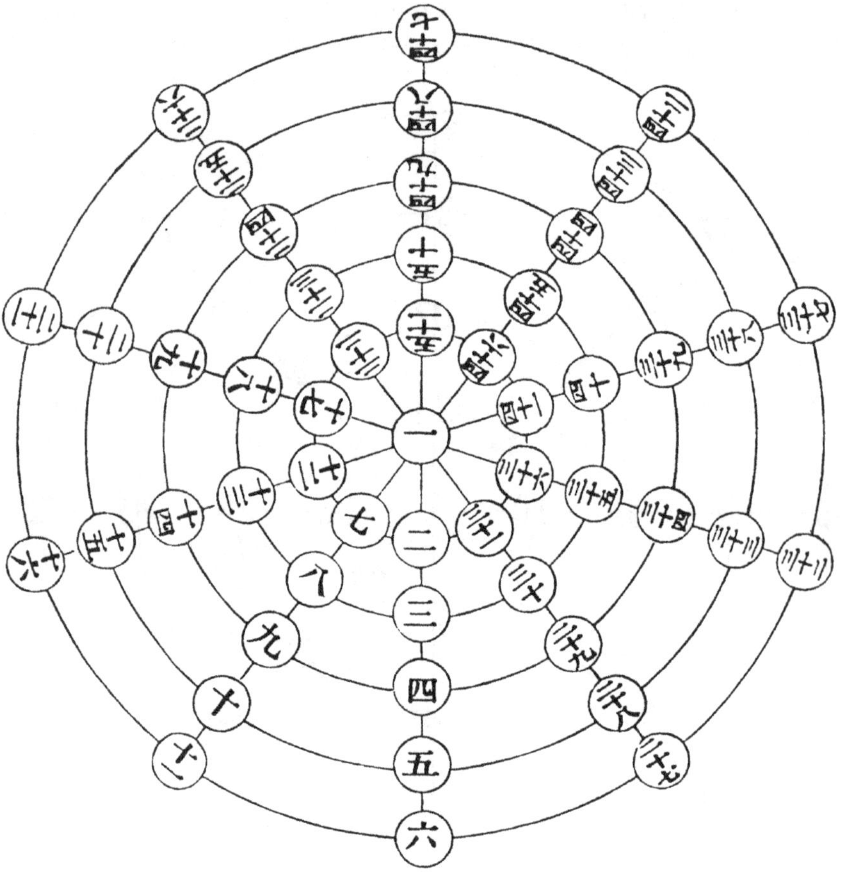

Fig. 29. Magic circle, from the Seki reprint of 1908.

the product of the cube of the chord and the moment of spherical volume.[2]

He finds this volume by taking from the sphere the central

---

[1] He calls it an "arc-ring," *kokan* or *kokwan* in Japanese.

[2] That is, the volume of a unit sphere. It is called by Seki the *ritsu-yen seki ritsu* or *gyoku seki hō*.

cylinder and the two caps.[1] He also considers the case in which the axis cuts the segment.

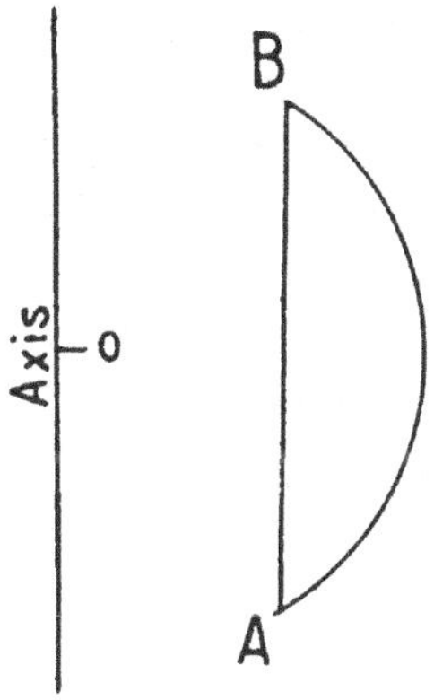

He likewise finds the volume generated by a lune formed by two arcs, the axis being parallel to the common chord, and either cutting the lune or lying wholly outside. Such work does not seem very difficult at present, but in Seki's time it was an advance over anything known in Japan.[2] These problems were to Japan what those of Cavalieri were to Europe, making a way for the *Katsujutsu* or method of multiple integration[3] of a later period.

Seki also concerned himself with indeterminate equations, beginning with $ax - by = 1$, to be solved for integers.[4] His first indeterminate problem is as follows: "There is a certain number of things of which it is only known that this number divided by 5 leaves a remainder 1, and divided by 7 leaves a remainder 2. Required the number."

---

[1] This is stated by an anonymous commentary known as the *Kyūketsu Hengyō Sō Genkai.*

[2] Endō, Book II, p. 45.

[3] Or rather the method of repeated application of the *tetsujutsu* expansion. Some of the problems involved only a single integration.

[4] This appears in his *Shūi Shoyaku no Hō,* written in 1683. His method of attacking these problems he calls the *senkan jutsu.* Problems of this nature appeared in the *Kwatsuyō Sampō.*

Since the number is evidently $5x+1$, and also $7y+2$, we have

$$5x+1 = 7y+2,$$

whence $$5x-7y=1,$$

which is in the form that he is considering. By what he calls the "method of leaving unity", he solves and finds that $x=3$, $y=2$, and the number is 16. He then proceeds to generalize the case for any number of divisors.[1]

Seki also gives the following typical problem:

"There is a certain number of things of which it is only known that this number multiplied by 35 and divided by 42 leaves a remainder 35; and multiplied by 44 and divided by 32 leaves a remainder 28; and multiplied by 45 and divided by 50 leaves a remainder 35. Required the number." His result is 13, and it is obtained by a plan analogous to the one used in the first problem. His other indeterminate problems show a good deal of ingenuity in arranging the conditions, but it is not necessary to enter further into this field.

One of the most marked proofs of Seki's genius is seen in his anticipation of the notion of determinants.[2] The school of Seki offered in succession five diplomas, representing various degrees of efficiency. The diploma of the third class was called the *Fukudai-menkyo*, and represented eighteen or nineteen subjects. The last of these subjects related to the *fukudai* problems or problems involving determinants, and since it appears in a revision of 1683,[3] its discovery antedates this year. Leibnitz (1646—1716), to whom the Western world generally assigns the first idea of determinants[4], simply asserted

---

[1] *Jō-ichi jutsu.* He seems to have taken it from the Chinese method of Ch'in Chiu-shao as set forth in the *Su-shu Chiu-chang* of 1247.

[2] T. HAYASHI, *The "Fukudai" and Determinants in Japanese Mathematics.* Tōkyō *Sūgaku-Buturigakkwai Kizi*, vol. V (2), p. 254 (1910).

[3] The *Fukudai-wo-kaisuru-hō* or *Kai-fukudai-no-hō* (Method of solving *fukudai* problems).

[4] T. MUIR, *Theory of Determinants in the historic order of its development.* London, 1890; D. E. SMITH, *History of Modern Mathematics.* New York, 1906, p. 26.

that in order that the equations

$$10 + 11x + 12y = 0, \quad 20 + 21x + 22y = 0, \quad 30 + 31x + 32y = 0$$

may have the same roots the expression

$$10.21.32 - 10.22.31 - 11.20.32 + 11.22.30 + 12.20.31 - 12.21.30$$

must vanish.[1] On the other hand, Seki treats of n equations. While Leibnitz's discovery was made in 1693 and was not published until after his death, it is evident that Seki was not only the discoverer but that he had a much broader idea than that of his great German contemporary. To show the essential features of his method we may first suppose that we have two equations of the second degree,

$$ax^2 + bx + c = 0$$
$$a'x^2 + b'x + c' = 0.$$

Eliminating x^2 we have

$$(a'b - ab')x + (a'c - ac') = 0,$$

and eliminating the absolute term and suppressing the factor x we have

$$(ac' - a'c)x + (bc' - b'c) = 0.$$

That is, we have two equations of the second degree and transform them into two equations of the first degree by what the Japanese called the process of folding (*tatamu*). In the same way we may transform n equations of the nth degree into n equations of the $n-1$ degree.[2] From these latter equations the *wasanka*[3] proceeded to eliminate the various powers of x. Since it was their custom to write only the coefficients, including all zero coefficients, and not to equate to zero,[4] their array of coefficients formed in itself a determinant, although they did not look upon it as a special function of the coefficients. On this array Seki now proceeds to per-

---

1 See MUIR, *loc. cit.*, p. 5.

2 Called *Kwanshiki* (substitute equations).

3 Follower of the *wasan* (native mathematics).

4 The second member always being zero in a Japanese equation.

form two operations, the *san* (to cut) and the *chi* (to manage). The *san* consisted in the removal of a constant literal factor in any row or column, exactly as we remove a factor from a determinant today. If the array (our determinant) equalled zero, this factor was at once dropped. The *chi* was the same operation with respect to a numerical factor.

Seki also expands this array of coefficients, practically the determinant that is the eliminant of the equations. In this expansion some of the products are positive and these are called *sei* (kept alive), while others are negative and are called *koku* (put to death), and rules for determining these signs are given. Seki knew that the number of terms in the expansion of a determinant of the nth order was $n!$, and he also knew the law of interchange of columns and rows.[1] Whatever, therefore, may be our opinion as to Seki's originality in the *yenri*,[2] or even as to his knowledge of that system at all or as to its value, we are compelled to recognize that to him rather than to Leibnitz is due the first step in the theory which afterwards, chiefly under the influence of Cramer (1750) and Cauchy (1812), was developed into the theory of determinants.[3] The theory occupied the attention of members of the Seki school from time to time as several anonymous manuscripts assert,[4] but the fact that nothing was printed leads to the belief

---

1 The details of these laws as expressed by the *wasanka* of the Seki school have been made out with painstaking care by Professor Hayashi, and for them the reader is referred to his article.

2 See Chapter VIII.

3 The best source for the history of the subject in the West is Muir, *loc. cit.*

4 Professor Hayashi has several in his possession. An anonymous one that seems to have been written in the eighteenth century, entitled *Fukudai riu san ka yendan justsu*, is in the library of one of the authors (D. E. S.). A contemporary of Seki's, Izeki Chishin, published a work entitled *Sampō Hakki* in 1690, in which the subject of determinants is treated, and upwards of twenty other works on the subject are now known. It is strange that the Japanese made no practical use of the idea in connection with the solution of linear equations, and entirely forgot the theory in the later period of the *wasan*.

that the process long remained a secret. It must be said, however, that the Chinese and Japanese method of writing a set of simultaneous equations was such that it is rather remarkable that no predecessor of Seki's discovered the idea of the determinant.

We have now considered all of Seki's work save only the mysterious *yenri*, or circle principle. It must be confessed that aside from his anticipation of determinants the result is disappointing. In Chapter VIII we shall consider the *yenri*, of which there is grave doubt that Seki was the author, and aside from this and his discovery of determinants his reputation has no basis in any great field of mathematics. That he was a wonderful teacher there can be no doubt; that he did a great deal to awaken Japan to realize her power in learning no one will question; that he was ingenious in improving mathematical devices is evident in everything he attempted; but that he was a great mathematician, the discoverer of any epoch-making theory, a genius of the highest order, there is not the slightest evidence. He may be compared with Christian Wolf rather than Leibnitz, and with Barrow rather than Newton. When, on November 15, 1907, His Majesty the Emperor of Japan paid great honor to his memory by bestowing upon him posthumously the junior class of the fourth Court rank, he rendered unprecedented distinction to a great scholar and a great teacher, but not to a great discoverer of mathematical theory.

CHAPTER VII.

Seki's contemporaries and possible Western influences.

Whether or not Seki can be called a great genius in mathematics, certain it is that his contemporaries looked upon him as such, and that he reacted upon them in such way as to arouse among the scholars of his day the highest degree of enthusiasm. Although he followed in the footsteps of Pythagoras in his relations with his pupils, admitting only a few select initiates to a knowledge of his discoveries,[1] and although he kept his discoveries from the masses and gave no heed to the researches of his contemporaries, nevertheless the fact that he could accomplish results, that he could solve the puzzling problems of the day, and that he had such a large following of disciples, made him a stimulating example to others who were not at all in touch with him. In view of this fact it is now proposed to speak of some of Seki's contemporaries before considering his own relation to the *yenri*, and at the same time to consider the question of possible Western influence at this period.

Two years before Seki published (1674) his *Hatsubi Sampō*, namely in 1672, Hoshino Sanenobu published his *Kokōgen-shō*, and in 1674 Murase, a pupil of Isomura, wrote the *Sampō Futsudan Kai*. A year later (1675), Yuasa Tokushi, a pupil of Muramatsu, published in Japan the Chinese *Suan-fa Tung-tsong*. In 1681 Okuda Yūyeki, a Nara physician, wrote the *Shimpen Sansū-ki*. Two years later, Takebe Kenkō published

---

[1] A custom always followed in the native Japanese schools, not merely in mathematics but also in other lines.

the *Kenki Sampō*, in which he solved the problems proposed in Ikeda Shōi's *Sūgaku Jōjo Ōrai* of 1672, without making use of the *tenzan* algebra of Seki, saying that "this touches upon what my mathematical master wishes kept secret," thus leaving unsolved those problems that required the *senkan-jutsu* and similar devices. It was in the work of Ikeda that the old Chinese value of π, $\frac{355}{113}$, was first made known in Japan.

In the same year (1683) Kozaka Sadanao published his *Kūichi Sangaku-sho.*[1] He had been the pupil of a certain Tokuhisa Kōmatsu, founder of the Kūichi school of mathematics, a school that was much given to astrology and mysticism.[2] Also in this year Nakanishi Seikō published his *Kōkogen Tekitō-shū*, a book that was followed in 1684 by the *Sampō Zoku Tekitō-shū* written by his brother, Nakanishi Seiri. These brothers had been pupils of Ikeda Shōi, and one of them[3] opened a school called after his name.

In 1684 the second edition of Isomura's *Ketsugi-shō* appeared,[4] and in the following year Takebe's commentary on Seki's *Hatsubi Sampō* was published. This latter made generally known the *yendan* method as taught by Seki.

In 1687 Mochinaga and Ōhashi published the *Kaisan-ki Kōmoku*,[5] and in 1688 the *Tōsho Kaisanki.*[6] In the first of these works we already find approaches to the crude methods of integration (see Fig. 30) that characterized the labors of the early Seki school. In the year 1688 Miyagi Seikō, the teacher of Ōhashi, published the *Meigen Sampō*, to be followed in 1695 by his *Wakan Sampō*[7] in which he considers in detail the numerical equation of the 1458th degree already mentioned by Seki, and attempts to solve the hundred fifty problems

---

1 Literally, the Mathematical Treatise of the Kūichi School.

2 Endō, Book II, p. 18.

3 The eldest, Nakanishi Seikō, may have studied under one of Seki's pupils. Endō, Book II, p. 20.

4 See p. 65.

5 Literally, the Summary of *Kaisan-ki*.

6 Literally, the *Kaisan-ki* with Commentary.

7 Japanese and Chinese Mathematical Methods.

in Satō's *Kongenki* and the fifteen in Sawaguchi's *Kokon Sampō-ki* (1670), all by the *yendan* process.

Miyagi founded a school in Kyōto that bore his name, and to him is sometimes referred a manuscript[1] on the quadrature of the circle. He was highly esteemed as a scholar by his contemporaries.[2]

In 1689 Andō Kichiji of Kyōto published a work entitled *Ikkyoku Sampō* in which the *yendan* algebra is set forth, and

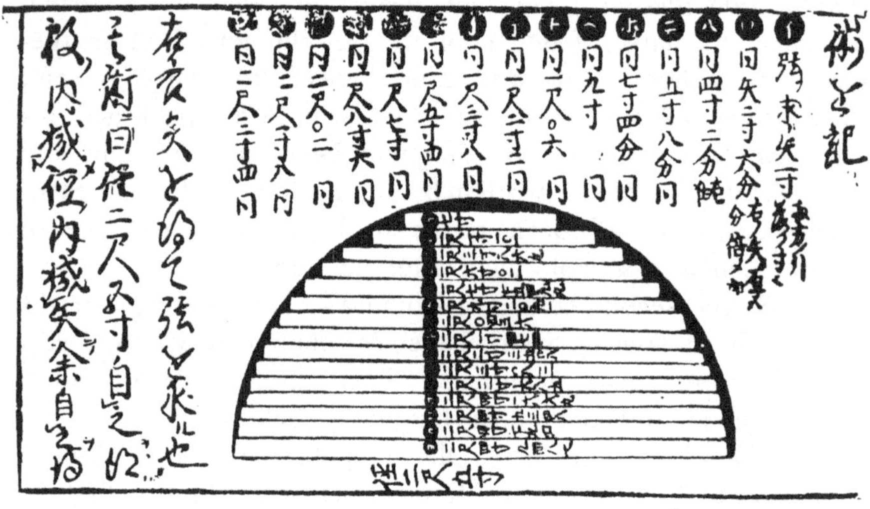

Fig. 30. Early integration, from Mochinaga and Ōhashi's *Kaisan-ki Kōmoku* (1687).

in 1691 Nakane Genkei published a sequel to it under the title *Shichijō Beki Yenshiki*.

In 1696, Ikeda Shōi published a pamphlet on the mensuration of the circle and sphere,[3] and in 1698 Satō Moshun

[1] The *Kohai Shōkai*. This is, however, an anonymous work of the eighteenth century.

[2] Endō, Book II, p. 29.

[3] The *Gyokuyen Kyoku-seki*, the Limiting Values of the circular Area and spherical Volume. In the same year (1696) Nakane Genkei published his *Tenmon Zukwai Hakki*, an astronomical work of importance. The best astronomical treatise of this period is Shibukawa Shunkai's *Tenmon Keitō*, a manuscript in 8 vols. Nakane Genkei also wrote a work on the calendar, the *Kōwa Tsūreki* that was later revised by Kitai Oshima.

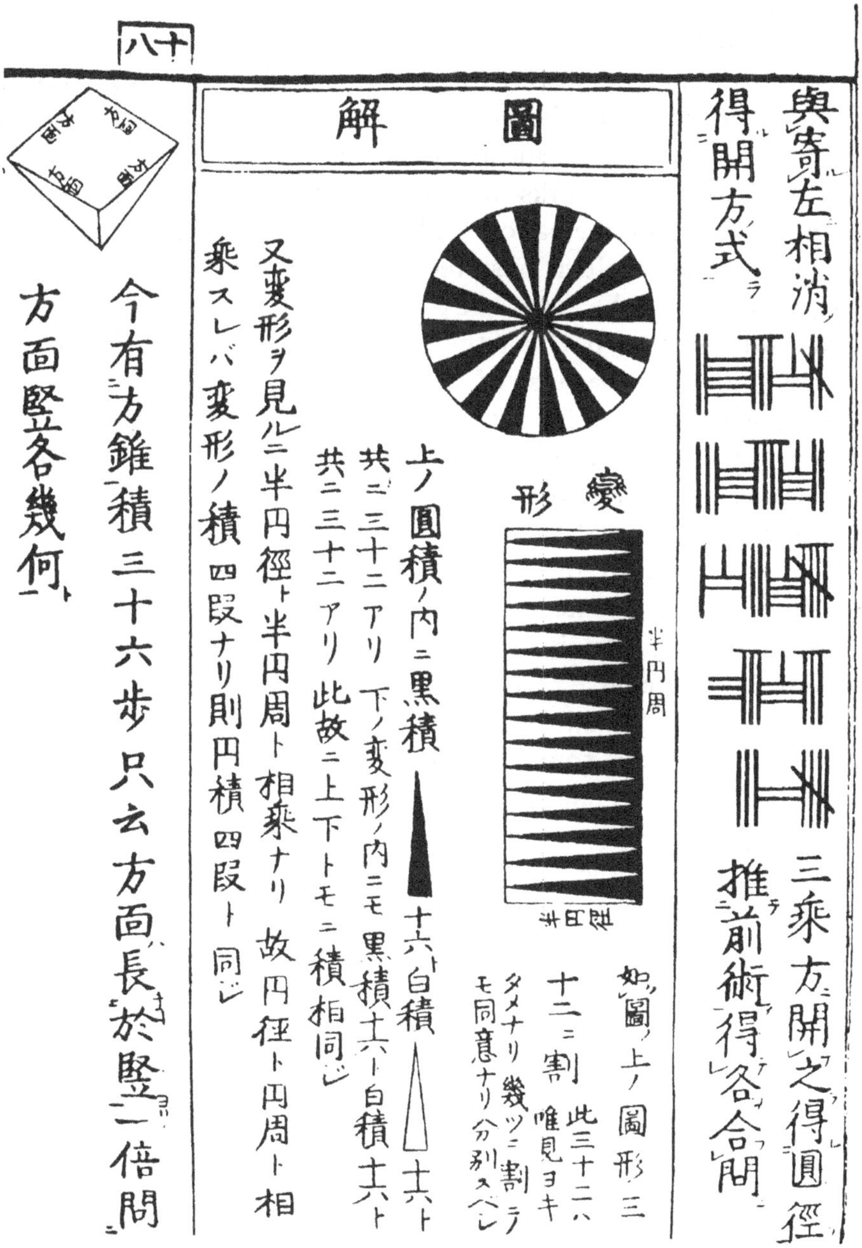

Fig. 31. Mensuration of the circle, from Satō Moshun's *Tengen Shinan* (1698).

published his *Tengen Shinan* or Treatise on the Celestial Element Method. In this his method of finding the area of a circle is distinctly Western (Fig. 31), although it is so simple as to claim no particular habitat.

This list is rather meaningless in itself, without further description of the works and a statement of their influence upon Japanese mathematics, and hence it may be thought to be of no value. It is inserted, however, for two purposes: first, that it might be seen that the Seki period, whether through Seki's influence or not, whether through the incipient influx of Western ideas or because of a spontaneous national awakening, was a period of special activity; and second, that it might be shown that out of a considerable list of contemporary writers, only those who in some way came under Seki's influence attained to any great prominence.

We now turn to the second and more important question, did Seki and his contemporaries receive an impetus from the West? Did the Dutch traders, who had a monopoly of the legitimate intercourse with mercantile Japan, carry to the scholars of Nagasaki and vicinity, where the Dutch were permitted to trade, some knowledge of the great advance in mathematics then taking place in the countries of Europe? Did the Jesuit missionaries in China, who had followed Matteo Ricci in fostering the study of mathematics in Peking, succeed in transmitting some inkling of their knowledge across the China Sea? Or did some adventurous scholar from Japan risk death at the order of the Shogun,[1] and venture westward in some trading ship bound homewards to the Netherlands? These are some of the questions that arise, and which there are legitimate reasons for asking, but they are questions that future research will have more definitely to answer. Some material for a reply exists, however, and the little knowledge that we have may properly be mentioned as a basis for future investigation.

It has for some time been known, for instance, that there

[1] Even the importation of foreign books was suppressed in 1630.

was a Japanese student of mathematics in Holland during Seki's time,[1] doubtless escaping by means of one of the Dutch trading vessels from Nagasaki. We know nothing of his Japanese name, but the Latin form adopted by him was Petrus Hartsingius, and we know that he studied under Van Schooten at Leyden. That he was a scholar of some distinction is seen in the fact that Van Schooten makes mention of him in his *Tractatus de concinnandis demonstrationibus geometricis ex calculo algebraico* in one of his editions of Descartes's *La Géométrie*,[2] as follows: "placuit majoris certitudinis ergo idem Theorema Syntheticé verificare, procendo à concessis ad quaesita, prout ad hoc me instigavit praestantessimus ac undequaque doctissimus juvenis D. Petrus Hartsingius, Iaponensis, quondam in addiscendis Mathematis, discipulus meus solertissimus."[3] The passage in Van Schooten was first noticed by Giovanni Vacca, who communicated it to Professor Moritz Cantor.

Some further light upon the matter is thrown by a record in the *Album Studiosorum Academiae Lugduno Batavae*,[4] as follows:

"Petrus Hartsingius Japonensis, 31, M. Hon. C." with the date May 6, 1669. Here the numeral stands for the age of the student, M. for medicine, his major subject, and Hon. C. for *Honoris Causa,* his record having been an honorable one.

---

1 Harzer, P., *Die exacten Wissenschaften im alten Japan, Jahresbericht der deutschen Mathematiker-Vereinigung*, Bd. 14, 1905, Heft 6; Mikami, Y., *Zur Frage abendländischer Einflüsse auf die japanische Mathematik am Ende des siebzehnten Jahrhunderts, Bibliotheca Mathematica*, Bd. VII (3), Heft 4.

2 Harzer quotes from the 1661 edition, p. 413. We have quoted from the Amsterdam edition of 1683, p. 413.

3 T. Hayashi remarks that the same words appear in a posthumous work of Van Schooten's, but this probably refers to the above *editio tertia* of 1683. See Hayashi, T., *On the Japanese who was in Europe about the middle of the seventeenth century* (in Japanese), *Journal of the Tokyō Physics School,* May, 1905; Mikami, Y., *Hatono Sōha and the mathematics of Seki,* in the *Nieuw Archief voor Wiskunde,* tweede Reeks, Negende Deel, 1910.

4 Hague, 1875. It gives a list of students and professors from 1575 to 1875.

Mathematics, his first pursuit, had therefore given place to medicine, and in this subject, as in the other, he had done noteworthy work. Possibly the death of Van Schooten in 1661 may have influenced this change, but it is more likely that the common union of mathematics and medicine, as indeed of all the sciences in those days,[1] led him to combine his two interests. Moreover certain other records inform us that Hartsingius lived in the house of one Pieter van Nieucasteel by the Langebrugge, a bit of information that adds a touch of reality to the picture. This record would therefore lead to the belief that he was only twenty-two years old when he was mentioned in the year of Van Schooten's death (1661), or probably only twenty-one when he, a *doctissimus juvenis*, and *quondam in addiscendis*, verified the theorem for his teacher.

A careful examination of the Leyden records as set forth in the *Album Studiosorum* throws a good deal more light on the matter than has as yet appeared. In the first place the Hartsingius was adopted as a good Dutch name, it appearing in such various forms as Hartsing and Hartsinck, and may very likely have belonged to the merchant under whose auspices the unknown student went to Holland. In the next place, Hartsingius was in Holland for a long time, fifteen years at least, and was off and on studying in the university at Leyden. He is first entered on the rolls under date August 29, 1654, as "Petrus Hartsing Japonensis. 20, P," a boy of twenty in the faculty of philosophy. This would have placed his birth in 1634 or 1635, but as we shall see, he was not very particular as to exactness in giving his age.[2] He next appears on the rolls in the entry of date August 28, 1660, "Petrus Hartzing Japonensis, 22, M." He has now changed his course to medicine, and his age would now place his birth in 1638 or 1639, four years later than stated before. Since, however,

---

1 Witness, for example, the mention made by Van Schooten in the 1683 edition (p. 385) above cited, of the assistence received from Erasmius Bartholinus, mathematician and physician in Copenhagen.

2 See *Album*, col. 438.

the difficulty of language is to be considered, together with the fact that such records, hastily made, are apt to be inexact, this is easily understood. He next appears in the *Album* under date May 6, 1669, as already stated. He therefore began in 1654, and was still at work in 1669, but he had not been there continuously.

Further light is thrown upon his career by the fact that he was not alone in leaving Japan, perhaps about 1652. He had with him a companion of the same age and of similar tastes. In the *Album*, under date September 4, 1654, appears this entry: "Franciscus Carron Japonensis, 20, P." Within a week, therefore, of the first enrollment of Hartsingius, another Japanese of same age, and doubtless his companion in travel, registered in the same faculty. But while Hartsingius remained in Leyden for years, we hear no more of Carron. Did he die, leaving his companion alone in this strange land? Did he go to some other university? Or did he make his way back to Japan?[1]

Now who was this Petrus Hartsingius who not only braved death by leaving his country at a time when such an act was equivalent to high treason, but who was excellent as a mathematician? What ever became of him? Did he die, an unknown though promising student, in some part of the West, or did he surreptitiously find his way back to his native land? If he passed his days in Europe did he send any messages from time to time to his friends, telling them of the great world in which he dwelt, and in particular of the medical work and the mathematics of the intellectual center of Northern Europe? In other words, for our immediate purposes, could the mathematics of the West, or any intimation of what was being accomplished by its devotees, have reached Japan in Seki's time?

---

[1] SCHOTEL, G. D. J., *De Academie te Leiden in de 16e, 17e en 18e eeuw* Haarlem, 1875, speaks (p. 266) of Japanese students at Leyden, and a further search may yield more information. We have been over the lists with much care from 1650 to 1670, and less carefully for a few years preceding and following these dates.

These questions are more easily asked than answered, but it is by no means improbable that the answers will come in due time. We have only recently had the problem stated, and the search for the solution has little more than just begun, while among all of the literature and traditions of the Japanese people it is not only possible but probable that the future will reveal that for which we are seeking.

At present there is a single possible clue to the solution. We know that a certain physician named Hatono Sōha, who flourished in the second half of the seventeenth century, did study abroad and did return to his native land.[1] Hatono was a member of the Nakashima[2] family, and before he went abroad he was known as Nakashima Chōzaburō. The family was of the *samurai* class, and formerly had been retainers of the Lord of Chōshū or of the Lord of Iwakuni,[3] feudal nobles who had made the Nakashimas at one time abundantly wealthy, but who had dishonestly deprived them of much of their means during the infancy of two of the heirs. It was because of this wrong that the family had left their former home and service and had repaired to the island of Kyūshū to seek to mend their fortunes. It was thus that they came to Nagasaki, and that the young Nakashima Chōzaburō met a Dutch trader with whom he departed into the forbidden world beyond the boundaries of the empire. It would seem, now, that we ought to be able to ascertain the date of the departure of the young

---

[1] For much of this information we are indebted to S. Hatono, a lineal descendent of the physician in question, and bearing his name. He informs us that the story was originally recorded in a manuscript entitled *Tsuboi Idan* which was destroyed by fire. See also Ishigami, T., *Hatono Sōha O* in the *Chūgwai Iji Shimpō*, no. 369, Aug. 5, 1895; Yokoyama, T., *A physician of the Dutch school who went abroad two centuries ago, and his surgical instruments* (in Japanese), in the *Kyōyuku Gakujutsu Kai*, vol. 4, January 1901, (an article that leaves much to be desired in the matter of clearness); Fujikawa, Y., *History of Japanese Medicine* (in Japanese); Yokoyama, T., *History of Education in Japan* (in Japanese).

[2] In the eastern part of Japan this name commonly appears as Nakajima, but Nakashima is the preferred form.

[3] The latter was subject to the former.

samurai, and to trace his wanderings, especially as he returned and could, at least in the secrecy of his family, have told his story. We are, however, quite uncertain as to any of these matters. His descendants have kept the tradition that his visit abroad was in the Manji era, and since this extended from 1658 to 1661, it included the time that Hartsingius was in Leyden. Tradition also says that he visited the capital of Namban, which at that time meant not only the Spanish peninsula, but the present and former colonies of Spain and Portugal, and which included Holland. While in this city he learned medicine from someone whose name resembled Postow or Bostow,[1] and after some years he again returned to Japan.

Arrived in his own country Nakashima was in danger of being beheaded for his violation of the law against emigration, and this may have caused the journeying from place to place which tradition relates of him. It is more probable, however, that his skill as a physician rendered him immune, the officials closing their eyes to a violation of the law which might be most helpful to themselves or their families in case of sickness. The danger seems to have passed through the permission granted by the Shogun that two European physicians, Almans and Caspar Schambergen should be permitted to practise at Nagasaki. Thereupon Nakashima became one of their pupils, began to practise in the same city, and assumed the name Nakashima Sōha.

It happened that there lived at that time in the province of Hizen, in Kyūshū, a certain daimyo who was very fond of a brood of pigeons that he owned. One of the pigeons having injured its leg, the daimyo sent for the young physician, and such was the skill shown by him, and so rapid was the recovery

---

[1] We have been unable to find this name among the list of prominent Spanish, Portuguse, or Dutch physicians of that time, but it is not improbable that some reader may identify it. Is it possible that it refers to Adolph Vorstius (Nov. 23, 1597—Oct. 9, 1663) who was on the medical faculty at Leyden from 1624 to 1663?

of the bird, that in all that region Nakashima's name became known and his praises were sung. So celebrated was his simple exploit that the people called him *Hato no ashi wo naoshita Sōha*[1] or *Hato no Sōha*,[2] a name so pleasing to him that he thereupon adopted it and was thenceforth known as Hatono Sōha.[3]

His fame now having found its way along the Inland Sea, a daimyo of the Higo province, Lord Hosokawa, in due time called him to enter his service at Ōsaka, so that he left Nagasaki, bearing with him gifts from his masters, Almans and Schambergen, as well as those which Postow had presented when he was in Europe or in some colony of Spain, Portugal, or Holland. This was in 1681,[4] and there he seems to have remained until his death in 1697, at the age of fifty-six years. Such is the brief story of the only Japanese scholar who is known, though native sources, to have studied in Europe and to have returned to his own country at about the time that Petrus Hartsingius was studying mathematics and medicine in Leyden. If Hatono was fifty-six when he died, as the family records assert, he must have been born in 1641 which is a little too late for Hartsingius, whereas if he and Carron are the same, his birth is placed in 1634 or 1635, which argues strongly against this conjecture.

The problem seems, therefore, to reduce to the search for a Doctor Postow, and to a search for some problem in the Japanese mathematics of the Seki school that is at the same time in Van Schooten's *Tractatus* or in some contemporary treatise. Thus far we have no knowledge that Hatono knew

---

1 Sōha who cured the pigeon's leg.

2 Sōha of the pigeon.

3 The name is now in the ninth generation.

4 This is the date as it appears in the family records, as communicated to us by his descendant. According to T. Yokoyama, however, there is a manuscript in the possession of the family, signed by Deshima Ranshyū at Nagasaki in 1684. If this is a *nom de plume* of Hatono's as Mr. Yokoyama believes, he may have gone to Ōsaka later than 1681.

any mathematics whatever.[1] If he was Hartsingius he could easily have communicated his knowledge to Seki or his disciples, and if he was not it is certain that he would have known him if he studied in Leyden, and in any case there is the mysterious Franciscus Carron to be considered.

As to Seki's contact with those who could have known the foreign learning, a story has long been told of his pilgrimage to the ancient city of Nara, then as now one of the most charming spots in all Japan, and still filled with evidence of its ancient culture. It appears that he had learned of certain treatises kept in one of the Buddhist temples, that had at one time been brought from China by the priests,[2] which related neither to religion nor to morals nor to the healing art, and which no one was able to understand. No sooner had he opened the volumes than he found, as he had anticipated, that they were treatises on Chinese mathematics, and these he copied, taking the results of his labor back to Yedo. It is further related that Seki spent three years in profitable study of these works, but what the books were or what he derived from them still remains a mystery.[3]

If Seki went to Nara, the great religious center of Japan, as there seems no reason to doubt, he would not have failed to visit the great intellectual center, Kyōto, which is near there. Neither would he have missed Ōsaka, also in the same vicinity, where Hatono Sōha was in the service of the daimyo. But

---

[1] Most of his manuscripts and the records of the family were burned some fifty years ago, and of the few that remained nearly all were destroyed at the siege of Kumamoto at the time of the Saigō rebellion in 1877.

[2] MIKAMI, Y., *On reading P. Harzer's paper on the mathematics in Japan, Jahresbericht der deutschen Math. Verein.*, Bd. XV, p. 256.

[3] Seki may have studied the Chinese work by Yang Hui at Nara. The story of his visit is said to have first appeared in the *Burin Inken Roku* or *Burin Kenbun Roku* written by one Saitō. It was reproduced in an anonymous manuscript entitled *Sanwa Zuihitsu,* possibly written by Furukawa Ken. It also appears in the *Okinagusa* written by Kamizawa Teikan. We have been unable to get any definite information as to the Nara books, although diligent inquiry has been made, but we wish to express our appreciation of the efforts in this direction made by Mrs. Kita (*née* Mayeda) and her brother.

on the other hand, Seki published the *Hatsubi Sampō* in 1674, while Hatono did not go to Ōsaka until 1681, so that in any event Seki could solve numerical equations of a high degree[1] before Hatono settled in his new home. Moreover the symbolism used by him is manifestly derived from the Chinese,[2] so that this part of his work shows no European influence. If Hatono or Hartsingius influenced Seki it must have been in the work in infinite series, which, as we shall see in the next chapter, started in his school, although more probably with his pupil Takebe.

Still another contact with the West is mentioned in a work called the *Nagasaki Semmin Den,* in which it is stated that one Seki Sōzaburō learned astronomy from an old scholar who had been to Macao and Luzon. If this is the Luzon of the Philippine Islands he could at that period have come in contact with the Jesuits, and this is very likely the case.

Mention should also be made of another possible medium of communication with the West in the time of Seki. Aside from the evident fact that if Hatono, Hartsingius, and Carron ventured forth on a voyage to Europe, others whose names are not now remembered may have done the same, we have the record of two men who were in touch with Western mathematics. These men were Hayashi Kichizaemon, and his disciple Kobayashi Yoshinobu, both of them interpreters in the open port of Nagasaki. Each of these men knew the Dutch language, and each was interested in the sciences, the latter being well versed in the astronomy of the West.[3] Kobayashi was suspected of being a convert to Christianity, and as this was a period of relentless persecution of the followers of this religion[4] he was thrown into prison in 1646, remaining there

---

1 He even hints at one of the 1458th degree (See page 129.)

2 Possibly obtained from Chinese works at Nara.

3 In 1650 a Portuguese whose Japanese name was Sawano Chūan wrote an astronomical work in Japanese, but in Latin characters. In 1659 Nishi Kichibei transliterated it and it was annotated by Mukai Gensho (1609—1677) under the title *Kenkon Bensetsu.*

4 It was in 1616 that the Tokugawa Shogunate ordered the strict sup-

for twenty-one years. Upon his release in 1667 he made an attempt to teach astronomy and the science of the calendar at Nagasaki,[1] though with what success is unknown, and it is recorded that in the year of his death, 1683, at the age of eighty-two, he was able to correct an error in the computation of an eclipse of the sun as recorded in the official calendar.[2] Hayashi was executed in 1646. While it is probable that these men did not know much of the European mathematics of the time, it is inconceivable that they were unaware of the general trend of the science, and that they should fail to give to inquirers some hint as to the nature of this work.

A little later than the time of Kobayashi there appeared still another scholar who knew the Dutch astronomy, one Nishikawa Joken, who was invited by the Shogun Yoshimune to compile the official calendar. As already stated, the latter was himself a dilletante in astronomy, and it was due to his foresight and to that of Nakane Genkei that the ban upon European books was raised in 1720. From this time on the astronomy of the West became well known in Japan, and scholars like Nagakubo Sekisui, Mayeno Ryōtaku, Shizuki Tadao, Asada Gōryū, and Takahashi Shiji were thoroughly acquainted with the works of the Dutch writers upon the subject.[3]

The conclusion appears from present evidence to be that some knowledge of European mathematics began to find its

---

pression of Christianity, the result being such a bloody persecution that a rebellion broke out at Shimabara, not far from Nagasaki, in 1637.

[1] ENDŌ, Book II, p. 76.

[2] ENDŌ, Book II, p. 18.

[3] Mayeno is said to have also had a Dutch arithmetic in 1772, but the title is not known. ENDŌ, Book III, p. 7. On this question of the influence of the Dutch see HAYASHI, T., *How have the Japanese used the Dutch books imported from Holland,* in the *Nieuw Archief voor Wiskunde,* reeks 2, deel 7, 1905, p. 42; 1906, p. 39, and later, where it appears that most of the Dutch works known in Japan are relatively late. On the interesting history of the Portuguese writer known as Sawano Chūan, see MIKAMI, Y., in the *Nieuw Archief voor Wiskunde,* reeks 2, deel 10, and the *Annals Scientificos da Academia Polytechnica do Porto,* vol. 7.

way into Japan in the seventeenth century; that we have no definite information as to the nature of this work beyond the fact that mathematical astronomy was part of it; that there is no evidence that Seki or his school borrowed their methods from the West; but that Japanese mathematicians of that time might very well have known the general trend of the science and the general nature of the results attained in European countries.

CHAPTER VIII.

The Yenri or Circle Principle.

Having considered the contributions of Seki concerning which there can be no reasonable doubt, and having touched upon the question of Western influence,[1] we now propose to examine the *yenri* with which his name is less positively connected. The word may be translated "circle principle" or "circle theory", the name being derived from the fact that the mensuration of the circle is the first subject that it treats. It may have been suggested by the title of the Chinese work of Li Yeh (1248), the *Tsê-yüan Hai-ching*, in which, as we have seen (page 49), *Tsî-yüan* means "to measure the circle." Seki himself never wrote upon it so far as is positively known, although tradition has assigned its discovery to him, nor is it treated by Ōtaka Yūshō in his *Kwatsuyō Sampō* of 1712 in connection with the analytic measurement of the circle. After Seki's time there were numerous works treating of the mere numerical measurement of the circle, such as the *Taisei Sankyō*,[2] commonly supposed to have been written by Takebe Kenkō,[3] and of which twenty books have come down to us out of a possible forty-three.[4] There is a story, generally considered as fabulous, told of three other books besides the twenty that are known, that were in possession of Mogami Tokunai[5] a century ago.

[1] The influence of the missionaries is considered later.

[2] "Complete Mathematical Treatise."

[3] So stated in a manuscript of Lord Arima's *Hōyen Kikō*, bearing date 1766.

[4] So stated by Oyamada Yosei in his article on the *Sangaku Shuban* in the *Matsunoya Hikki*, although the number is doubtful.

[5] A pupil of Honda Rimei (1755—1836).

He stated that he procured them from one Shiono Kōteki of Hachiōji, who had learned mathematics from Someya Harufusa. Shiono recorded these facts at the end of his copy, and this is the bearing of the story upon Seki's secret knowledge of the *yenri*. It was Someya who gave Shiono these books, assuring him that they contained Seki's secret knowledge, being works that he had himself written. Someya had received them from Ishigaya Shōyeki of Kurozawa in Sagami, his aged master, who was a pupil of Seki's and who had received these copies from the latter's own hand.

Although the story is not a new one, and seems to relate Seki intimately with the work, nevertheless we have no evidence save tradition to corroborate the statement, since the three volumes no longer exist, if they ever did, and the twenty that we know show no evidence of being Seki's work.[1] Moreover the treatment of π which it contains is quite certainly not that of Seki, for in his *Fukyū Tetsujutsu* of 1722 Takebe states that it is not.[2] This treatment is based upon the squares of the perimeters of regular inscribed polygons from 4 to 512, π^2 being taken as the square of the perimeter of the 512-gon, namely

$$9.86960\ 44010\ 89358\ 61883\ 44901\ 99874\ 7.$$

Seki, on the contrary, calculated the successive perimeters instead of their squares. Takebe claims to have carried his process far enough to give π to upwards of forty decimal places by considering only a 1024-gon, and he gives it as

$$\pi = 3.14159\ 26535\ 89793\ 23843\ 26433\ 83279\ 50288\ 41971\ 2.$$[3]

He then uses continued fractions to express this value, stating that this plan is due to his brother Takebe Kemmei, and that

---

[1] It should be stated, however, that ENDŌ (Book II, p. 41) believes, and with excellent reason, that they were taken from Seki's own writings and were put into readable form by Takebe. See also MIKAMI, Y., *A Question on Seki's Invention of the Circle-Principle*, in the *Tōkyō Sūgaku-Buturigakkwai Kizi*, Book IV (2), no. 22, p. 442, and also his article on the *yenri* in Book V (2).

[2] MS., article 10.

[3] He must, however, have gone beyond the 1024-gon for this.

Seki had used only the method given in the *Kwatsuyō Sampō*, all of which tends to throw doubt upon Seki's connection with this treatise.

The successive fractions obtained for π by taking the convergents of the continued fraction are

$$\frac{3}{1},\quad \frac{22}{7},\quad \frac{333}{106},\quad \frac{355}{113},$$

$$\frac{103993}{33102},\quad \frac{104348}{33215},\quad \frac{208341}{66317},\quad \frac{312689}{99532},$$

$$\frac{833719}{265381},\quad \frac{1146408}{364913},\quad \frac{4272943}{1360120},\quad \text{etc.},$$

most of which are not found in any work with which we can clearly connect Seki's name.

Still another reason for doubting Seki's relation to this phase of the work is seen in the method of measuring a circular arc. In the *Taisei Sankyō* the squares of the arcs are used instead of the arcs themselves, as in the case of the circle. Some idea of the work of this period may be obtained from the formula given:

$$(48773156 87c^6 + 21309475994c^4h^2 + 23945445808c^2h^4 + 5170741462h^6)a^2$$
$$= 48775156 87c^8 + 47322893653c^6h^2 + 151469740022c^4h^4 + 174277533560c^2h^6 + 50319088000h^8,$$

where c = chord, h = height of arc (from the center of the chord to the center of the arc),[1] and a = length of arc. This formula resembles one that appears in the *Kwatsuyō Sampō*, and one that is in Takebe's *Kenki Sampō* of 1683. All these formulas seem due to Seki.

Some idea of the *Taisei Sankyō* having been given, together with some reasons for doubting the relation of Seki to it, we shall now speak of the author, Takebe, and of his other works, and of his use of the *yenri*, setting forth his testimony as to any possible relation of Seki to the method.

---

[1] Which we shall hereafter call the height of the arc, the older word *sagitta* being no longer in common use.

Takebe Hikojirō Kenkō[1] was one of three brothers who displayed a taste for mathematics[2] and who studied under Seki. He was descended from an ancient family, his father Takebe Chokukō being a shogunate *samurai*. He was born in Yedo (Tōkyō) in the sixth month of 1664, and while still a youth became a pupil of Seki, and, as it turned out, his favorite and most distinguished one.[3]

Takebe was only nineteen years of age when he published the *Kenki Sampō* (1683). Two years later (1685) there appeared his commentary on Seki's *Hatsubi Sampō* (of 1674), and in 1690 he wrote the seven books of his notes on the *Suan-hsiao Chi-mêng* which appeared in his edition of this work,[4] explaining the *sangi* method of solving numerical equations. In 1703 he was made a shogunate *samurai* and served as an official in the department of ceremonies. In 1719 he drew a map of Japan, upon which he had been working for four years, and which for its accuracy and for the delicacy of his work was looked upon as a remarkable achievement. This and his vast range of scientific knowledge served to command the admiration and respect of Yoshimune, the eighth of the Tokugawa shoguns, who called upon him for advice with respect to the calendar and who consulted him upon matters relating to astronomy, a subject in which each took a deep interest. He at once pointed out certain errors in the official calendar, and recommended as court astronomer Nakane Genkei, for whom and for himself Yoshimune built an observatory in

---

1 His given name Kenkō appears as Katahiro in the *Hakuseki Shinsho* written by Arai Hakuseki (1657—1725), his contemporary, and is so given in some of the histories. It is possible too that the family name Takebe should be Tatebe, as given by Endō, Okamoto, and others of the old Japanese school, although the former is usually given.

2 The other brothers were his seniors and were called Kenshi and Kemmei, also known as Katayuki and Kataaki.

3 Kawakita, C., *Honchō Sūgaku Shiryō* (Materials for the Mathematical History of Japan), pp. 63—66, this being based upon Furukawa Ujikiyo's writings. See also Kuichi Sanjin's article in the *Sūgaku Hōchi*.

4 This Chinese algebra appeared in 1299. The Japanese edition is mentioned in Chapter IV.

the castle where he dwelt. So liberal minded was this shogun that he removed the prohibition upon the importation of foreign treatises upon medicine and astronomy, so that from this time on the science of the West was no longer under the ban.

The infirmities of age began to tell upon Takebe in 1733 so much as to lead him to resign his official position, and six years later, on the twentieth day of the seventh month of the year 1739, he passed away at the age of seventy-five years.

The work of Takebe's with which we are chiefly concerned was written in 1722, and was entitled *Fukyū Tetsujutsu*, Fukyū being his *nom de plume*, and Tetsujutsu being the Japanese form of the title of a Chinese work written by Tsu Ch'ung-chi (430—501) in the fifth century. This Chinese work is now lost, but it treated of the mensuration of the circle,[1] and for this reason there is an added interest in the use of its name in a work upon the *yenri*.

Takebe states[2] that Seki was wont to say that calculations relating to the circle were so difficult that there could be no general method of attack. Indeed he says that Seki was averse to complicated theories, while he himself took such delight in minute analysis that he finally succeeded in his efforts at the quadrature of the circle. It would thus appear that the *yenri* was not the product of Seki's thought, but rather of Takebe's painstaking labor. Moreover the plan followed by Takebe in finding the length of an arc is not the same as the one given in the *Kwatsuyō Sampō* in which Ōtaka Yūshō (1712) sets forth Seki's methods, though it has some resemblance to that given in the *Taisei Sankyō* which, as we have seen, Takebe may have written in his younger days when he was more under Seki's influence.

---

[1] As we know from Wei Chi's Records of the Sui Dynasty, a work written in the seventh century. It was possibly a treatise on the calendar in which the circle was considered incidentally. See Mikami, Y., in the *Proceedings of the Tōkyō Math. Phys. Society*, October, 1910.

[2] Article 8 of his treatise.

Takebe takes a circle of diameter 10 and finds the square of half an arc of height 0.000001 to be a number expressed in our decimal system as

0.00000 00000 33333 35111 11225 39690 66667 28234 77694 79595 875 +,

but he gives us no complete explanation as to how this was obtained.[1] Now since the squares of the halves of arcs of heights 1, 0.1, and 0.00001, respectively, have for their approximate values 10, 1, and 0.0001, it will be observed that these are the products of the diameter and the heights of the arcs. He therefore takes dh, the product of the diameter and height, as the first approximation to the square of half an arc. He then compares this approximation with the ascertained value and takes his first difference D_1 as $\frac{1}{3}h^2$. Proceeding in a similar manner he finds the second difference D_2 to be $\frac{h}{d}\cdot\frac{8}{15}\cdot D_1$, and so on for the successive differences. The result is the formula

$$\frac{1}{4}a^2 = dh + \frac{1}{3}h^2 + \frac{h}{d}\cdot\frac{8}{15}\cdot D_1 + \frac{h}{d}\cdot\frac{9}{14}\cdot D_2 + \frac{h}{d}\cdot\frac{32}{45}\cdot D_3 + \frac{h}{d}\cdot\frac{25}{33}\cdot D_4 + \frac{h}{d}\cdot\frac{72}{91}\cdot D_5 + \cdots$$

In other words, he has

$$\frac{1}{4}a^2 = dh\left[1 + \sum_1^\infty \frac{2^{2n+1}(n!)^2}{(2n+2)!}\cdot\left(\frac{h}{d}\right)^n\right],$$

which expresses in a series the square of arc sin x in terms of versin x.

This series is convenient enough when h is sufficiently small, but it is difficult to use when h is relatively large. Takebe

[1] He states that the particulars are set forth in two manuscripts, the *Yenritsu* (Calculation of the Circle) and *Koritsu* (Calculation of the Circular Arc), but these manuscripts are now lost.

therefore developed another series to be used in these cases, as follows:

$$\frac{1}{4}a^2 = dh + \frac{1}{3}h^2 + \frac{h}{d-h}\cdot\frac{8}{15}\cdot D_1 - \frac{h}{d-h}\cdot\frac{5}{14}\cdot D_2 + \frac{h}{d-h}\cdot\frac{12}{15}\cdot D_3 - \frac{h}{d-h}\cdot\frac{223}{398}\cdot D_4 \cdots$$

He also gives a third series which he, possibly following Seki, derives from the value of $h = 0.00000\ 0001$, as follows:

$$\frac{1}{4}a^2 = dh + \frac{1}{3}h^2 + \frac{8}{15}\cdot\frac{h}{d-\frac{9}{14}h}\cdot D_1 + \frac{43}{980}\cdot\frac{h^2}{d^2+\frac{6743008}{26176293}h^2-\frac{1696}{1419}dh}\cdot D_2 + \ldots.$$

Takebe's method of finding the surface of a sphere is the same as that given in the revised edition of Isomura's *Ketsugishō* save that it is carried to a closer degree of approximation. As bearing upon Seki's work it should be noted that Takebe states that the former disdained to follow this method, preferring to consider the center as the vertex of a cone of which the altitude equals the radius, showing again that Takebe was quite independent of his master.

Not only does Takebe use infinite series in the manner already shown, but in another of his works he does so in a still more interesting fashion. This work has come down to us in manuscript under the title *Yenri Tetsujutsu* or *Yenri Kohai-jutsu.*[1] In this he considers the following problem: In a segment of a circle the two chords of the semi-arc are drawn, after which arcs are continually bisected and chords are drawn. The altitude of half the given arc then satisfies the equation

$$-dh + 4dx - 4x^2 = 0,$$

where d = diameter, h = altitude of the given arc, x = altitude of half of this arc. This equation Takebe proceeds to solve

[1] Literally, The circle principle, or Method of finding the arc of a circle.

by expressing the value of x in the form of a series, expanded according to a process which he calls *Kijo Kyūshō jutsu.*[1]

From this expansion Takebe derives a general formula for the square of an arc, which he gives substantially as follows:

$$\frac{a^2}{4} = dh\left[1 + \sum_{1}^{\infty}\frac{2^2 \cdot 4^2 \ldots\ldots (2n)^2}{3 \cdot 4 \cdot 5 \cdot 6 \ldots (2n+1)(2n+2)}\left(\frac{h}{d}\right)^n\right]$$

$$= dh\left[1 + \sum_{1}^{\infty} 2^{2n+1}\frac{(n!)^2}{(2n+2)!}\left(\frac{h}{d}\right)^n\right],$$

a result that had previously been obtained in the *Fukyū Tetsujutsu* of 1722.[2]

The analysis leading to this formula, which is too long to be given here and which is obscure at best, is the *yenri* or Circle Principle, and it at once suggests two questions: (1) What is its value? (2) Who was its discoverer?

As to each of these questions the answer is difficult. In the first place, Takebe does not state with lucidity his train of reasoning, and we are unable to say how he bridged certain difficulties that seem to have stood in his way. He gives us results instead of a principle, an isolated formula instead of a powerful method. To be sure his formula has, as we shall see, some interesting applications, as have also many formulas of the calculus; but here is only one formula, obscurely derived, whereas the calculus is a theory from which an indefinite number of formulas may be derived by lucid reasoning. We are therefore constrained to say that, from any evidence offered by Takebe, the *yenri* is simply the interesting, ingenious, rather obscure method of deriving a formula capable of being applied in several ways, but that it is in no more comparable to the European calculus, even as it existed in the time of Seki, than is Archimedes's method of squaring the parabola, while the method is stated with none of the lucidity of the great Syracusan.

---

1 Literally, Method of deriving the root by divisions.

2 See page 148, above.

But taking it for what it is worth, who invented the *yenri?* The greatest of Japanese historians of mathematics, Endō, is positive that it was Seki. He sets forth the reasons for his belief as follows:[1] "The inventions of the *tenzan* algebra and of the *yenri* were made early [in the renaissance of Japanese mathematics], but certain scholars do not attribute the latter to Seki for the reason that it is not mentioned in the *Kwatsuyō Sampō*. Such a view of the question is, however, entirely unwarranted. At that period even the *tenzan* algebra was kept a profound secret in Seki's school, never being revealed to the uninitiated. It was on this account that not even the *tenzan* algebra was treated in the *Kwatsuyō Sampō*, and hence there is little cause for wonder that the *yenri* has no place there. It is stated, however, that the value of π is slightly less than 3.14159265359. Now unless the correct value were known [to this number of decimal places] how would this fact have been evident? . . . The process given in this work being restricted to the inscription of polygons, there was no means of knowing how many digits are correct. Nevertheless the author was correct in his statement as to how many decimal places are exact, and so it would seem that he must already have known the correct value to more decimal places [than were used] in order to make his comparison. The original source of information was certainly one of Seki's writings, perhaps the same as that used by Takebe in his subsequent work."

While Endō's argument thus far is not conclusive, since Seki may have found the value of π by the older process, or may have obtained it from the West, nevertheless it must be granted that, as Takebe assures us, he did know it to more than twenty figures.

Endō continues: "In the Kyōhō era (1716—1736) Seki's adopted son, Shinshichi, was dismissed from office and was forced to live under Takebe's care. It was at this juncture that Takebe, in consultation with him, entered upon a study

---

[2] Endō, Book II, pp. 55, 56.

of Seki's most secret writing on the *yenri* as applied to the rectification of a circular arc, after which he completed his manuscript entitled *Yenri Kohai Tetsujutsu*".[1] He continues[2] by saying that Shinshichi was dismissed from office in the Shogunate in 1735 because of his dissolute character, so that we thus have a date which will serve as a limit for such communication as may have taken place. He asserts that Seki's adopted son now gave to Takebe the secret writings of his father, written in the Genroku era (1688—1704) or earlier, and it was through their study that Takebe came to elaborate the *yenri*. Endō thinks that Takebe did not enter upon this work before the dismissal of Seki's adopted son in 1735 at which time he was already an old man.[3]

Now it is evident that this view of the case is not wholly correct, for Takebe gives the same series in his *Fukyū Tetsujutsu* in 1722. Moreover, he must have been acquainted with that form of analysis because there is extant a manuscript compiled in 1728 by one Ōyama (or Awayama) Shōkei[4] entitled *Yenri Hakki* which is quite like the *Yenri Kohai-jutsu* in its main features, although the work is not so minutely carried out, in spite of its gain in simplicity.

For example, the square of the arc is given in a series which is substantially the same as the one already assigned to Takebe. Ōyama's rule may be put in modern form as follows:

$$a^2 = 4\,dh\left[1 + \sum_{1}^{\infty} \frac{2^{2n+1}(n!)^2}{(2n+2)!}\left(\frac{h}{d}\right)^n\right].$$

From this series he derives the value of π by writing $h = \frac{d}{2}$

1 Endō, Book II, p. 74.

2 Ibid., pp. 81, 82.

3 His reasons are not clear. Professor T. Hayashi, in his article in the *Honchō Sūgaku Kōenshu*, 1908, pp. 33—36, makes out a strong case for Seki as the discoverer of the *yenri*.

4 Possibly Tanzan Skōkei. The writer of the preface of the work, Hachiya Teishō, may have been this same person.

and taking four times the result. He also finds it by taking $h = d$, the result being

$$\pi^2 = 4\left[1 + \sum_{1}^{\infty} \frac{2^{2n+1}(n!)^2}{(2n+2)!}\right].$$

Ōyama, the author of the *Yenri Hakki*, was a pupil of Kuru Jūson, who had studied under Seki, but the theory is not given as in any way connected with the latter. In one of the two prefaces Nakane Genkei, a pupil of Takebe's, says: "The most difficult problem having to do with numbers is the quadrature of the circle. On this account it is that we have the various results of the different mathematicians. . . . It is now a century since the dawn of learning in our country, and during this period divers discoveries have been made. Of these the most remarkable one is that of Takebe of Yedo. For several decades he has pursued his studies with such zeal that oftimes he has forgotten his need of food and sleep. In the spring of 1722 he was at last rewarded by brilliant success, for then it was that he came upon the long-sought formula for the circle. Since then he has shown his result to divers scholars, all of whom were struck with amazement, and all of whom cried out, 'Human or divine! This drives away the clouds and darkness and leaves only the blue sky!' And so it may be said that he is the one man in a thousand years, the light of the Land of the Rising Sun!"

The second preface is by Hachiya Kojūrō Teishō, and he too gives the credit to Takebe. He says, "The circle principle is a perfect method, never before known in ancient or in modern times. It is a method that is eternal and unchangeable . . . It is the true method, constructed first by the genius of Takebe Kenkō, and before him anticipated neither in Japan nor in China. It is so wonderful that Takebe should have made such a valuable discovery that it is only natural to look upon him as divine. For years have I studied under Seki's pupil Kuru Jūson, and have labored long upon the problem of the quadrature of the circle, but only of late have I learned of

Takebe's discovery, and I shall be happy if this work, which I have written, may initiate my fellow mathematicians into the mysteries of the problem."

It would seem from the last sentence that Hachiya may have been the real author of the work, and that Ōyama Shōkei and Hachiya may have been the same person. In any case, however, the evidence is clear that his contemporaries proclaimed Takebe the discoverer of the *yenri*, and there seems to have been none to challenge this award. There is no contemporary statement like this that connects the principle with Seki, and until there is stronger evidence than mere conjecture such honor as is due should be bestowed upon Takebe.

But where did Takebe get this formula for a^2? His explanation of his own development is very obscure. Did he himself understand it, or had he the formula and did he explain it as far as his ingenuity allowed? That there is a close resemblance between this formula and such series as one finds in looking over the works of Wallis[1] is evident. The series seems, however, to have been given by Pierre Jartoux, a Jesuit missionary, resident in Peking. This Jartoux was born in 1670 and went to China in 1700, dying there Nov. 30, 1720. He was a man of all-round intelligence,[2] and his *Observations astronomiques*, published two years after his death, showed some ability. He also worked with Père Régis on the great map of China. But our interest in Jartoux lies chiefly in the fact that he was in correspondence with Leibnitz, as is shown by the publication

[1] Our attention is called to this fact by P. HARZER, *Die exakten Wissenschaften im alten Japan*, in the *Jahresbericht der deutschen Mathemat. Verein.*, Bd. 14, Heft 6. A search through Wallis fails, however, to reveal this series, although the analogy to this work is evident. See, for example, WALLIS, J., *De Algebra Tractatus*, Oxoniae, 1693, cap. XCVI. The attention of readers is invited to the desirability of ascertaining if this series was already known in Europe.

[2] His report, *Détails sur le Ging-seng, et sur la récolté de cette plante*, published in Europe in 1720, was the best one upon the subject that had appeared in the West up to that time. Indeed it is for this report that he was best known there.

of his *Observationes Macularum Solarium Pekino missae ad G. W. Leibnitium* in the *Acta Eruditorum*.[1]

Here then is a scholar, Jartoux, in correspondence with Leibnitz, giving a series not difficult of deduction by the calculus, which series Takebe uses and which is the essence of the *yenri*, but which Takebe has difficulty in explaining, and which he might easily have learned through that intercourse of scholars that is never entirely closed. There is a tradition that Jartoux gave nine series,[2] of which three were transmitted to Japan,[3] and it seems a reasonable conjecture that Western learning was responsible for his work, that he was responsible for Takebe's series, and that Takebe explained the series as best he could.

The knowledge of Takebe's work was the signal for the appearance of various treatises upon the *yenri* besides that of Ōyama, and while they add nothing of importance to the theory or to its history, mention should be made of a few. The one that was the most highly esteemed in the Seki school of mathematicians was the *Kenkon no Maki*,[4] a work of unknown authorship.[5] Not only is the author unknown, but the work itself is apparently no longer extant in its original form.[6] The

---

[1] In 1705, p. 485.

[2] Professor Hayashi thinks that Jartoux did not give nine series, but that he gave six, and that these were obtained by Ming An-tu whose work was completed by his pupils after his death, and published in 1774. Among these six is Takebe's series. *Proceedings of the Tōkyō Math. Phys. Soc.*, 1910 (in Japanese).

[3] These three appear in Mei Ku-cheng's book, but the date is unknown and there is no evidence that it reached Japan in this period.

[4] Literally, The Rolls of Heaven and Earth.

[5] Endō thinks that it was written by Matsunaga; see his *History*, Book II, p. 84. P. Harzer thinks the author was Yamaji; see the *Jahresbericht der deutschen Morgenl. Ver.*, Bd. 14, p. 317. C. Kawakita thinks it was Araki, and in Fukuda's *Sampō Tamatebako* (1879) the same opinion is expressed.

[6] A manuscript bearing this title was found in a private library at Sendai, in the possession of a former pupil of Yamaji, but N. Okamoto, who has investigated the matter, believes that it is quite different from the original treatise.

process followed in developing the formula for a^2 is simpler than that used by Takebe in his *Yenri Kohai-jutsu* and rather resembles that of Ōyama.

The unknown author finds that the altitudes for the successive arcs formed by doubling the number of chords are

$$h_1 = \frac{1}{4}h\left[1 + \frac{1}{4}\left(\frac{h}{d}\right) + \frac{1.3}{4.6}\left(\frac{h}{d}\right)^2 + \frac{1.3.5}{4.6.8}\left(\frac{h}{d}\right)^3 + \cdots\right],$$

$$h_2 = \frac{1}{16}h\left[1 + \frac{5}{16}\left(\frac{h}{d}\right) + \frac{5.21}{16.40}\left(\frac{h}{d}\right)^2 + \frac{5.21.143}{16.40.224}\left(\frac{h}{d}\right)^3 + \cdots\right],$$

$$h_3 = \frac{1}{64}h\left[1 + \frac{21}{64}\left(\frac{h}{d}\right) + \frac{4.17}{64.32}\left(\frac{h}{d}\right)^2 + \frac{4\ 17.575}{64.32.896}\left(\frac{h}{d}\right)^3 + \cdots\right],$$

these being calculated by the *tetsujutsu* process, or the actual expansion of the terms of the equations, although the calculations themselves are not given. The ratios of the successive coefficients are seen to be

$$\frac{1.3}{3.4},\ \frac{3.5}{5.6},\ \frac{5.7}{7.8},\ \frac{7.9}{9.10},\ \frac{9.11}{11.12},\ \frac{11.13}{13\ 14},\ \frac{13.15}{15.16},\ \ldots,$$

$$\frac{3.5}{6.8},\ \frac{7.9}{10.12},\ \frac{11.13}{14.16},\ \frac{15.17}{18.20},\ \frac{19.21}{22.24},\ \frac{23.25}{26.28},\ \frac{27.29}{30.32},\ \ldots,$$

$$\frac{7.9}{12.16},\ \frac{15.17}{20.24},\ \frac{23.25}{28.32},\ \frac{31.33}{36.40},\ \frac{39.41}{44.48},\ \frac{47.49}{52.56},\ \frac{55.57}{60.64},\ \ldots.$$

Hence the mth ratio for h_r is of the form

$$\frac{(km-1)\,(km+1)}{(km+\frac{1}{2}k)\,(km+k)} = \frac{2\,(k^2m^2-1)}{k^2\,(2m^2+3m+1)}$$

where $k = 2^r$, and as k becomes infinite this reduces to

$$\frac{2m^2}{2m^2+3m+1}.$$

We therefore have the limit to which h is approaching, and we can compute the square of the arc as before. This is the plan as stated in the Sendai manuscript, the only one which it seems safe to use, even though the manuscript is evidently not like the lost original.[1]

[1] ENDŌ, Book II, pp. 84—90, gives a different treatment, resembling that found in the *Kohai no Ri*. None of the leading mathematicians of the

There is some little testimony in favor of Seki's authorship of the *Kenkon no Maki,* although the presumption is entirely against it. Thus in an anonymous work entitled *Kigenkai* or *Yenri Kenkon Sho,* a note by Furukawa Ujikiyo relates the following: "This book is a writing of Seki Kōwa and has long been kept a profound secret. No one into whose hands it has come was entitled to assume the rôle of Seki's successor. Hence Fujita Sadasuke treasured the work, and copied it upon two rolls which he called *Kenkon no Maki,*[1] revealing it only to his son and to his most celebrated pupil. All this has been told me by Shiraishi Chōchū." The probabilities are that some parts of the work were simply an ancient paraphrase of Ōtaka Yūshō's *Kwatsuyō Sampō,* and being thus of the Seki school it was attributed to the master. Whether or not it was the original *Kenkon no Maki* is unknown. However that may be, it extends the *yenri* to include the analytic treatment of the volume of a spherical segment of one base of diameter a, by a method not unlike that of Cavalieri. The segment is divided into n thin layers of diameters $d_1, d_2, \ldots d_n$, where $d_n = a$. Then

$$d_k^2 = 4\left(d - \frac{kh}{n}\right)\frac{kh}{n},$$

where d = diameter of the sphere, and h = altitude of the segment. Summing for $k = 1, 2, 3, \ldots n$, we have

$$\sum_1^n d_k^2 = 4\frac{dh}{n}\sum_1^n k - \frac{4h^2}{n^2}\sum_1^n k^2$$

$$= \frac{4dh}{n}\cdot\frac{n+n^2}{2} - \frac{4h^2}{n^2}\cdot\frac{n+3n^2+2n^3}{6}.$$

Multiplying this by $\frac{h}{n}$ and by $\frac{\pi}{4}$, we have the approximate volume of the spherical segment,

latter part of the nineteenth century received the *Kenkon no Maki* (possibly another name for the *Kohai no Ri*) from their teachers, as Uchida Gokan told N. OKAMOTO and as we are assured by T. HAGIWARA.

[1] See page 155, note 4.

$$\frac{\pi h^2}{6}\left(3d - 2h - \frac{h}{n^2} - \frac{3h}{n} + \frac{3d}{n}\right),$$

of which the limit for $n = \infty$ is

$$\frac{\pi h^2}{6}(3d - 2h).$$

The same general method appears in the writings of Matsunaga, Yamaji, and others.

It has already been stated that Isomura and Takebe found the spherical surface by means of the difference of volumes of two concentric spheres. In this work the same thing is done for the surface of an ellipsoid. The volume of the solid is given as $\frac{\pi ab^2}{6}$, but with no proof. Another ellipsoid is taken with axes $a + 2k$ and $b + 2k$, and the difference of their volumes is divided by k, giving

$$\frac{\pi}{3}(2ab + b^2 + 2ak + 4bk + 4k^2),$$

the limit of which, for $k = 0$, is

$$\frac{\pi}{3}(2ab + b^2).$$

This treatment is an improvement upon that of Isomura and Takebe because it is general rather than numerical. We therefore have here a further development of the *yenri*, in which it takes on a little more of the nature of the Western calculus, but still in only a narrow fashion.

In the same way, little by little, some progress was made in the use of infinite series. Takebe's series for the circular arc appears again in 1739 in a work entitled *Hōyen Sankyō*,[1] written by Matsunaga Ryōhitsu,[2] who received the secrets of the Seki school from Araki, under whom he had studied. The Araki-Matsunaga school, while it started under a less brilliant leader than the school of Takebe, became the more prosperous

---

1 Literally, Mathematical Treatise on Polygons and Circles.

2 His former name was Terauchi Gompei. He is also known as Matsunaga Yoshisuke.

as time went on, and seems to have inherited most of Seki's manuscripts. Araki, indeed, gave the name to Seki's Seven Books,[1] and upon his death in 1718,[2] at the age of seventy-eight, he could look back upon intimate associations with the mathematics of the past, and upon the renaissance in the labors of Seki, and could anticipate a fruitful future in the promise of Matsunaga.

Matsunaga was born at Kurume in Kyūshū, or possibly in Terauchi in Awari. His given name being Terauchi Gompei, we find some of his works signed with the name Terauchi. He served under Naitō Masaki, Lord of Taira in Iwaki and afterward Lord of Nobeoka in Kyūshū, himself no mean mathematician. Indeed it was he whose insistence led Matsunaga to adopt the name *tenzan* for the Japanese algebra, replacing the name *Kigen seihō* as used by Seki. Matsunaga was a prolific writer[3] and it is to him that the perpetuation of the doctrines of the master, under the title "School of Seki", was due. He died in the sixth month of 1744.[4]

In the statutes of the school of Seki, as laid down by him, the work was arranged in five classes, Seki himself having arranged it in three. The two upper classes were termed *Betsuden* and *Inka*,[5] the latter covering Seki's Seven Books, and being open only to one son of the head of the school and to two of the most promising pupils. These three initiates were required to take a blood oath of secrecy,[6] and still further

---

[1] The *Seki-ryū Shichibusho*, published at Tokyō as a memorial volume on the two hundredth anniversary of Seki's death. See also Endō, Book II, p. 42. There is some doubt as to the titles of the seven books.

[2] C. Kawakita in the *Honchō Sūgaku Koenshū*, p. 1.

[3] His works include the following: *Darui Shōsa* (1716), *Embi Empi Ryōjutsu* (1735), *Hōrō Yosan*, *Hōyen Sankyō* (1739), *Hōyen Zassan*, *Kaikō Un-ō* (1747, posthumous), *Kijo Tokushō*, *Sampō Shūsei*, *Sampō Tetsujutsu*.

[4] As stated in a manuscript by Hagiwara.

[5] These names may possibly mean "Special Instruction" and "Revealed by Swearing." One who completed these classes received the two diplomas known as *Betsuden-menkyo* and *Inka-menkyo*.

[6] Endō, Book II, p. 82 seq. On the five diplomas see also Hayashi, T., *The Fukudai and Determinants in Japanese Mathematics*, in the *Tōkyō Sūgaku-*

analogy to the ancient Pythagorean brotherhood is seen in the mysticism of the founder. Matsunaga writes[1] as Pythagoras might have done: "Reason is determinate, but Spirit wanders in the realm of change. Where Reason dwelleth, there is Number found; and wheresoever Spirit wanders, there Number journeys also. Spirit liveth, but Reason and Number are inanimate, and act not of their own accord. The way whereby we attain to Number is called The Art. Heaven is independent, but wherever there are things there is Number. Things, Number,—these are found in nature. What oppresses the high and exalts the humble; what takes from the strong and gives to the weak; what causes plenty here and a void there; what shortens that which is long and lengthens that which is short; what averages up the excess with the defect,—this is the eternal law of Nature. All arts come from Nature, and by the Will alone they cannot exist."

Matsunaga's *Hōyen Sankyō* is composed of five books, and is devoted entirely to formulas for the circumference and arcs of a circle, no analyses appearing.[2] His first series is as follows:

$$\frac{\pi^2}{9} - 1 + \frac{1^2}{3 \cdot 4} + \frac{1^2 \cdot 2^2}{3 \cdot 4 \cdot 5 \cdot 6} + \frac{1^2 \cdot 2^2 \cdot 3^2}{3 \cdot 4 \cdot 5 \cdot 6 \cdot 7 \cdot 8} + \cdots$$

This is followed by

$$\frac{\pi}{3} = 1 + \frac{1^2}{4 \cdot 6} + \frac{1^2 \cdot 3^2}{4 \cdot 6 \cdot 8 \cdot 10} + \frac{1^2 \cdot 3^2 \cdot 5^2}{4 \cdot 6 \cdot 8 \cdot 10 \cdot 12 \cdot 14} + \cdots,$$

a series which is then employed for the evaluation of π to fifty figures. The result is the following:

$$\pi = 3.14159\ 26535\ 89793\ 23846\ 26433\ 83279\ 50288\ 41971\ 69399\ 5751.$$

Buturigakkwai Kizi, vol. V (2), no. 5, 1910. Yamaji seems to have revealed the secrets to three besides his son.

[1] *Hōyen Sankyō*, 1739. This work may have been closely connected with the anonymous *Kohai Shōkai.*

[2] We are informed by N. Okamoto that Uchida Gokan used to say that the original manuscripts containing the analyses were burned purposely after the work was finished. Matsunaga's *Hōyen Zassan* (Miscellany concerning Regular Polygons and the Circle) is now unknown.

The same value is given in the *Hōyen Kikō*, written by Lord Arima in 1766, together with the numerical calculations involved. The value was first actually printed in the *Shūki Sampō*, written by Arima under an assumed name, in 1769.

Matsunaga next gives Takebe's series for the square of an arc,[1] this being followed by three series for the length of an arc a with chord c as follows:

$$a = c\left[1 + \frac{2}{3}\left(\frac{h}{d}\right) + \frac{2.4}{3.5}\left(\frac{h}{d}\right)^2 + \frac{2.4.6}{3.5.7}\left(\frac{h}{d}\right)^3 + \cdots\right],$$

$$a = 2\sqrt{hd}\left[1 + \frac{1^2}{2.3}\left(\frac{h}{d}\right) + \frac{1^2.3^2}{2.3.4.5}\left(\frac{h}{d}\right)^2 + \frac{1^2.3^2.5^2}{7!}\left(\frac{h}{d}\right)^3 + \ldots\right],$$

$$a = \frac{4dh}{c}\left[1 - \frac{1}{3}\cdot\left(\frac{h}{d}\right) - \frac{2}{3.5}\left(\frac{h}{d}\right)^2 - \frac{2.4}{3.5.7}\left(\frac{h}{d}\right)^3 - \cdots\right].$$

The series for the altitude h in terms of the arc is

$$h = \frac{d}{2}\sum_{0}^{\infty}(-1)^n\frac{1}{(2n)!}\cdot\left(\frac{a}{d}\right)^{2n},$$

and for the chord c it is

$$c = a - \frac{a^3}{2.3d^2} + \frac{a^5}{2.3.4.5.d^4} - \frac{a^7}{2.3.4.5.6.7d^6} + \cdots,$$

which is at once seen to be a form of the series for sin a.[2]

The area s of a circular segment is given as

$$s = \frac{2}{3}ch\left[1 + \frac{1}{5}\left(\frac{h}{d}\right) + \frac{6}{5.7}\left(\frac{h}{d}\right)^2 + \frac{6.8}{5.7.9}\left(\frac{h}{d}\right)^3 + \cdots\right]^3$$

where c = chord of the arc, d = diameter of the circle, and h = height of the segment.

Matsunaga also gives some interesting formulas for computing the radius x of a circle circumscribed about a regular polygon of n sides, one side being s, and for computing the apothem.

---

[1] Which appeared in the *Yenri Kohai-jutsu* and the *Fukyū Tetsu-jutsu* of Takebe and the *Yenri Hakki* of Ōyama.

[2] These two series appear in the *Shūki Sampō*.

[3] The above series are given in the *Hōyen Sankyō*, Book I.

He also gives[1] formulas for the side of the inscribed polygon in terms of the diameter of the circle, for the various diagonals, for the lines joining the mid-points of the diagonals and the various vertices or the mid-points of the sides,[2] and so on, none of which it is worth while to consider in a work of this nature.

It will be seen that the *yenri* as laid down by Takebe was extended to include solid figures treated somewhat after the manner of Cavalieri, but that it was little more than a rather primitive method of using infinite series in the measurement of the simplest curvilinear figures and the sphere. We shall see, however, that it gradually unfolds into something more elaborate, but that it never becomes a great method, remaining always a set of ingenious devices.

---

1 *Hōyen Sankyō,* Book III.

2 Lines known as the *Kyomen-shi.*

CHAPTER IX.

The eighteenth Century.

We have already spoken of the closing labors of Seki Kōwa, who died in 1708, and of Takebe Kenkō and Araki, and in Chapter X we shall speak of Ajima Chokuyen. There were many others, however, who contributed to the progress of mathematics from the time when Takebe made the *yenri* known to the days when Ajima gave a new impulse to the science, and of these we shall speak in this chapter. Concerning some of them we know but little, and concerning certain others a brief mention of their works will suffice. Others there are, however, who may be said to have done a work that was to that of Seki what the work of D'Alembert and Euler was to that of Newton. That is to say, the periods in Japan and Europe were somewhat analogous in a relative way, although the breadth of the work in the two parts of the world was not on a par. In some respects the period immediately following Seki was, save as to Takebe's work, one of relative quiet, of the gathering up of the results that had been accomplished and of putting them into usable form, or of solving problems by the new methods. In the history of mathematics such a period usually and naturally follows an era of discovery.

So we have Nishiwaki Richyū publishing his *Sampō Tengen Roku* in 1714, setting forth in simple fashion the "celestial element" and the *yendan* algebra.[1] In 1722 Man-o Tokiharu published his *Kiku Buntō Shū*, in which he treated, among other topics, the spiral. In 1715 Hozumi Yoshin published his

---

[1] ENDŌ, Book II, pp. 57, 59.

Kagaku Sampō, the usual type of problem book. In 1716 Miyake Kenryū published a similar work, the *Guwō Sampō*. He also wrote the *Shojutsu Sangaku Zuye*, of which an edition appeared in 1795 (Fig. 32). In this he seems to have had some idea of the prismatoid (Fig. 33). In 1718 Ogino Nobutomo wrote a work, the *Kiku Gempo Chōken*, that has come down to us in nine books in manuscript form,—a very worthy

Fig. 32. From Miyake Kenryū's *Shojutsu Sangaku Zuye* (1795 edition).

general treatise. Inspired by Hozumi Yoshin's work, Aoyama Riyei published his *Chūgaku Sampō* in 1719, solving the problems of the *Kagaku Sampō* and proposing others. These latter were solved in turn by Nakane Genjun in his *Kantō Sampō* (1738), by Nakao Seisei in his *Sangaku Bemmō*, and by Iriye Shūkei in his *Tangen Sampō* (1759). Mention should also be made of an excellent work by Murai Mashahiro, the *Ryōchi Shinan*, of which the first part appeared in 1732. The work was a popular one and did much to arouse an interest

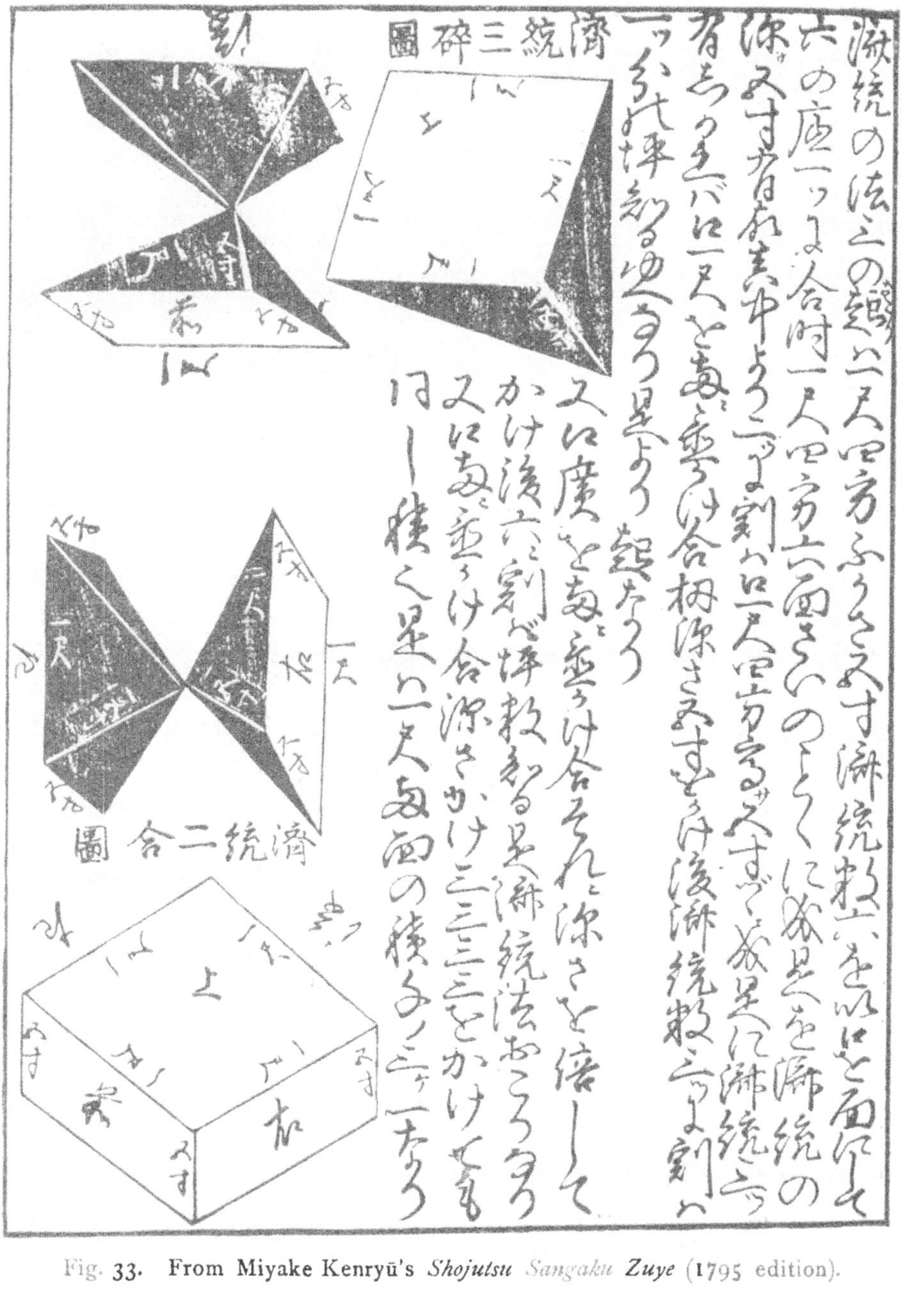

Fig. 33. From Miyake Kenryū's *Shojutsu Sangaku Zuye* (1795 edition).

in the new mathematics. The problems proposed by Nakane Genjun were answered by Kamiya Hōtei in his *Kaishō Sampō* (1743), by Yamamoto Kakuan in his *Sanzui*, and by others. To the same style of mathematics were devoted Yamamoto's *Yōkyoku Sampō* (1745) and *Keiroku Sampō* (1746), Takeda Saisei's *Sembi Sampō* (1746), Imai Kentei's *Meigen Sampō* (1764), and various other similar works, but by the close of the eighteenth century in Japan, as elsewhere, this style of book lost caste as representing a lower form of science than that in which the best type of mind found pleasure. Mention should also be made of Baba Nobutake's *Shogaku Tenmon* of 1706, a well-known work on astronomy, that exerted no little influence at this period (Fig. 34).

Of the writers of this general class one of the best was Nakane Genjun (1701—1761), whose *Kantō Sampō* (1738) attracted considerable attention. His father, Nakane Genkei (1661—1733), was born in the province of Ōmi, and studied under Takebe. He was at one time an office holder, but in earlier years he practiced as a physician at Kyōto. His taste led him to study mathematics and astronomy as well, and he seems to have been a worthy instructor for his son, who thus received at second hand the teachings of Seki's greatest pupil. Some interesting testimony to his standing as a scholar is given in a story related of a certain feudal lord of the Kyōhō period (1716—1736), who asked a savant, one Shinozaki, who were his most celebrated contemporaries. Thereupon the savant replied: "Of philosophers, the most celebrated are Itō Jinsai and Ogyū Sorai; of astronomers, Nakane Genkei and Kurushima Kinai;[1] in calligraphy, Hosoi Kōtaku and Tsuboi Yoshitomo; in Shintōism, Nashimoto of Komo; in poetry Matsuki Jiroyemon; and as an actor, Ichikawa Danjyūrō. Of these, Nakane is not only versed in astronomy, but he is eminent in all branches of learning."[2]

Nakane the Elder also published several astronomical works,

---

[1] Or Kurushima Yoshita.

[2] K. KANO's article in the *Honchō Sūgaku Kōenshū*, 1908, p. 11.

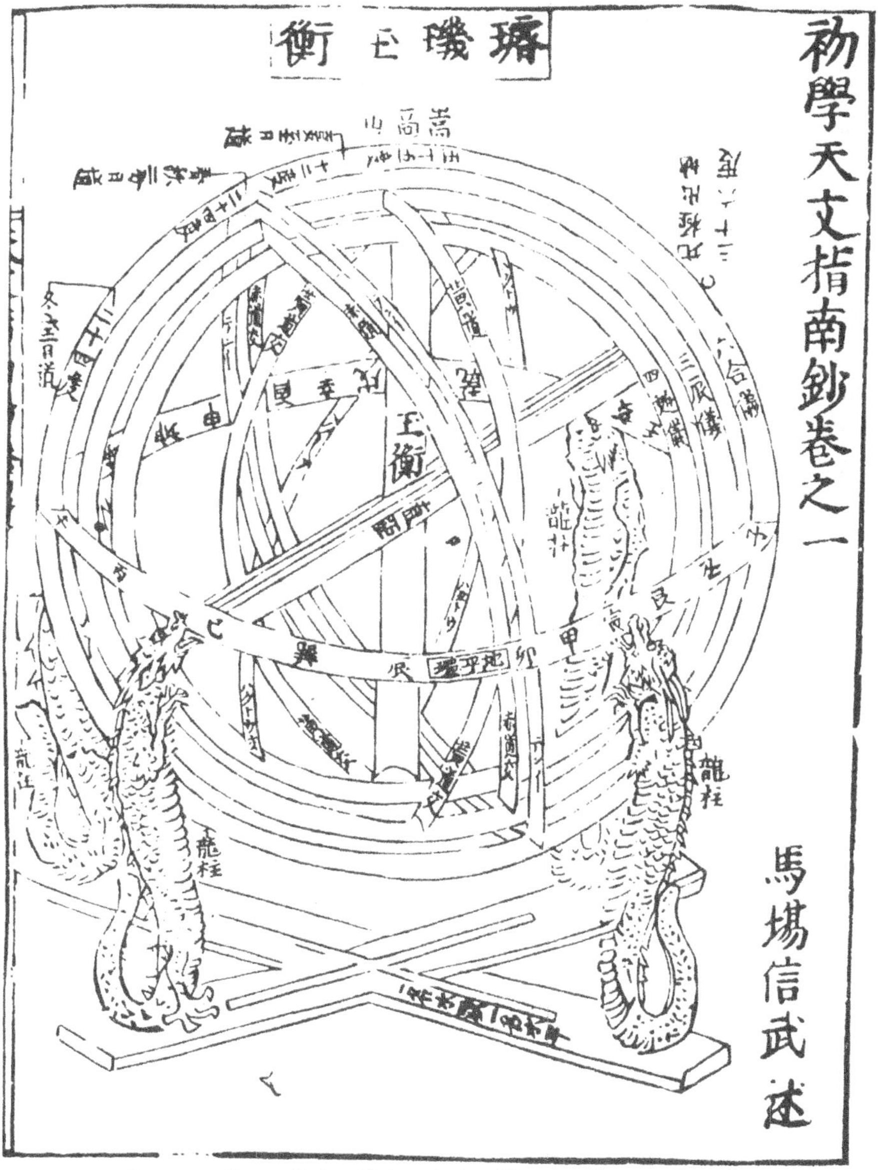

Fig. 34. From Baba Nobutake's *Shogaku Tenmon* (1706).

and composed a treatise in which a new law of musical melodies was set forth.[1] Through the Chinese works and the

[1] This was the *Ritsugen Hakki*, a work on the description of measures.

writings and translations of the Jesuit missionaries in China he was familiar with the European astronomy, and he recognized fully its superiority over the native Chinese theory. He was prominent among those who counseled the Shogun Yoshimune to remove the prohibition against the importation and study of foreign books, and by order of the latter he is said to have translated Mei Wen-ting's *Li-suan Ch'üan-shu.*[1] In 1711 he was given a post in the mint at Ōsaka, and in 1721 became connected with the preparation of the official calendar.[2] In pure mathematics he wrote but one work that was published, the *Shichijō Beki Yenshiki*,[3] although by all testimony he was an able mathematician. One of his solutions, appearing in Takebe's *Fukyū Tetsu-jutsu* (1722), is that of an interesting indeterminate equation. The problem is to find the sides of a triangle that shall have the values n, $n+1$, and $n+2$, and such that the perpendicular upon the longest side from the opposite vertex shall be rational. Nakane solves it as follows:

When the sides are 1, 2, 3, the perpendicular is evidently zero.

Taking the cases arising from increasing these values successively by unity, the following triangles satisfy the conditions:

3	13	51	193
4	14	52	194
5	15	53	195

If we represent these values by a_1, b_1, c_1; a_2, b_2, c_2; a_3, b_3, c_3; . . ., it will readily be seen that

$$a_{r+1} = 4a_r + 2 - a_{r-1},$$

and similarly for the b's and c's, and hence we have the required solution. Whether or not he made the induction complete does not, however, appear.

---

1 See page 19. The work is in the library of the Emperor.

2 For this purpose he spent half of his time in Yedo, the rest being spent in Kyōto.

3 It was printed in 1691 and reprinted in 1798.

It is also related that Takebe was asked in 1729, by the Shogun Yoshimune, for the solution of a certain problem on the calendar. Takebe, recognizing the great ability of the aged Nakane, asked him to undertake it; but he, feeling the infirmities of his years, passed it in turn to his son, Nakane Genjun. The result was a new method of solving numerical higher equations by successive approximations that alternately exceed and fall short of the real value, a method that was embodied in the *Kaihō Yeijiku-jutsu*[1] written by Nakane Genjun in 1729. The problem proposed by the Shōgun is as follows:[2] "There are two places, one in the south and one in the north, from which the elevation of the pole star above the horizon is 36° and 40°75′ respectively. At noon on the second day of the ninth month in a certain year the shadows of rods 0.8 of a yard high were 0.59 of a yard and 0.695 of a yard, respectively, and at the southern station the center of the sun was 36° 37′ distant from the zenith at noon on the day of the equinox. Required from these data to determine the ratio of the diameter of the sun's orbit to the diameter of the earth, considering the two to be concentric."

The solution of this problem is too long to be given here, but that of another one in the same manuscript may serve to illustrate Nakane's methods. "Given a circle in which are inscribed two equal smaller circles and another circle which we shall designate as the middle circle. Each of these four circles is tangent to the other three; the difference of area between the large circle and the three inscribed circles is 120, and the diameters of the middle and small circles differ by 5. Required to find the diameters."

Nakane lets l, m, s, stand for the respective diameters of the large circle, middle circle, and small circles.

Then $$s + 5 = m$$

and $$(s + m)^2 - s^2 = a^2,$$ an arbitrary abbreviation.

---

[1] Literally, Method of Increase and Decrease in the Evolution of Equations.

[2] From a manuscript of 1729.

Then $$l = \frac{(a+m)^2}{2(a+5)},$$

and $$l^2 - 2s^2 - m^2 = 102 : \frac{\pi}{4}.$$

He then assumes that $s_1 = 7.5$,

whence, from the above, the two sides of the equation become

150.0654 and 152.788,

their difference, d_1, being 2.723.

He next tries $$s_2 = 7.6,$$

whence, as before, $$d_2 = -0.37811.$$

He then takes $$s_3 = s_1 + \frac{d_1}{\frac{d_1 + d_2}{s_2 - s_1}} = 7.5878,$$

whence as before, $$-d_3 = -0.028246.$$

He now proceeds as before, taking

$$s_4 = s_2 - \frac{d_2}{\frac{d_2 - d_3}{s_2 - s_3}} = 7.5868\ldots,$$

and in the same way he continues his approximations as far as desired.

Not only did Nakane the younger study with his father, but he also went to Yedo (Tōkyō) to learn of Takebe and of Kurushima. Returning to Ōsaka he succeeded his father in the mint, and in 1738 he published the *Kantō Sampō* followed in 1741 by an arithmetic for beginners under the title *Kanja Otogi Zōshi*.[1] In this latter work the mercantile use of the *Soroban* is explained (Fig. 35) and the check by the casting out of nines is first used in multiplication, division, and evolution in Japan. He died in 1761 at the age of sixty.

The most distinguished of Nakane Genkei's pupils was Kōda Shin-yei, who excelled in astronomy rather than in pure

---

[1] Literally, A Companion Book for Arithmeticians.

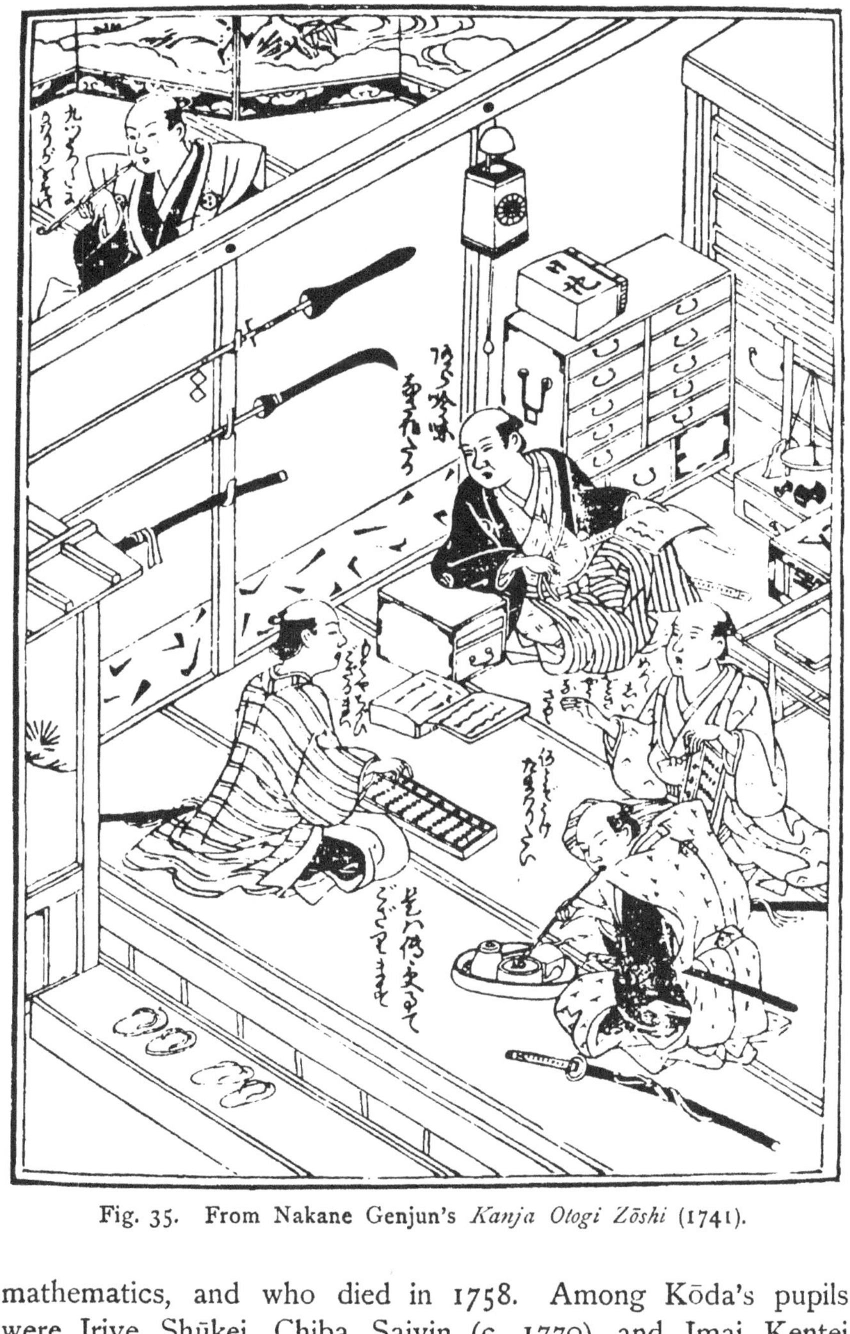

Fig. 35. From Nakane Genjun's *Kanja Otogi Zōshi* (1741).

mathematics, and who died in 1758. Among Kōda's pupils were Iriye Shūkei, Chiba Saiyin (c. 1770), and Imai Kentei (1718—1780). Imai Kentei, who left several unpublished manuscripts, had as his most prominent pupil Honda Rimei (1751—1828),[1] a man of wide learning and of great influence in education. Honda numbered among his pupils many distinguished men, including Aida Ammei, Murata Kōryū, Kusaka Sei, Mogami Tokunai, Sakabe Kōhan, and Baba Seitoku. He gave much attention to the science of navigation and to public affairs, and even advocated the opening of Japan to foreign trade. He was familiar with the Dutch language, and made some attempt at mathematical research,[2] and to his influence Mamiya Rinzō, the celebrated traveler, acknowledged his deep indebtedness.

Another prominent disciple of Takebe's was Koike Yūken (1683—1754), a *samurai* of Mito, where he presided over the *Shōkōkwan* or Institute for Historical Research. By order of his lord he went to Yedo and learned mathematics from Takebe, acquiring at the same time some knowledge of astronomy.

His successor in the *Shōkōkwan* at Mito was Ōba Keimei (1719—1785), but neither one contributed anything to mathematics beyond a sympathetic interest in the progress of the science.

Among the pupils of Nakane Genjun, and therefore of the Takebe branch of the Seki school, was Murai Chūzen, a Kyōto physician. He wrote a work entitled the *Kaishō Tempei Sampō*[3] (1765) which treated of the solution of numerical higher equations. Three years later one of his pupils, Nagano Seiyō, published a second part of this work in which he attempted to explain the methods employed in the solutions. For example, Murai[4] takes the equation

$$6726 - 373x + x^2 = 0.$$

He then finds the relation

$$373 - 372 \cdot 1 = 1,$$

---

[1] Also known as Honda Toshiaki.

[2] OZAWA, *Lineage of mathematicians* (in Japanese), and the epitaph on Honda's tomb.

[3] Literally, The Posting of Soldiers in the Evolution of Equations.

[4] ENDŌ, Book II, pp. 137—139.

and multiplies the 372 into the absolute term (6726) and then subtracts 373 as often as possible, leaving a remainder 361.[1] This remainder is added to 6726 and the result is divided by 373, the quotient, 19, being a root.

Similarly, in the equation

$$-25233 - 2284x + 25x^3 = 0,$$

Murai claims first to take the relation

$$2284 \times 11 - 25m = -1,$$

and states that he multiplies 11 into the absolute term, subtracting 2284 from the product until he reaches a remainder, which is the root required, a process that is not at all clear.

Of course the method is not valid, for in the equation

$$x^2 - 8x + 15 = 0$$

it gives 2 instead of 3 or 5 for the root. Murai must have been aware that his rule was good only for special cases, but

Fig. 36. From Murai Chūzen's *Sampō Dōshi-mon* (1781).

[1] Briefly, $372 \times 6726 = 2{,}502{,}072$, and $2{,}502{,}072 \div 373 = 6707$ with a remainder 361.

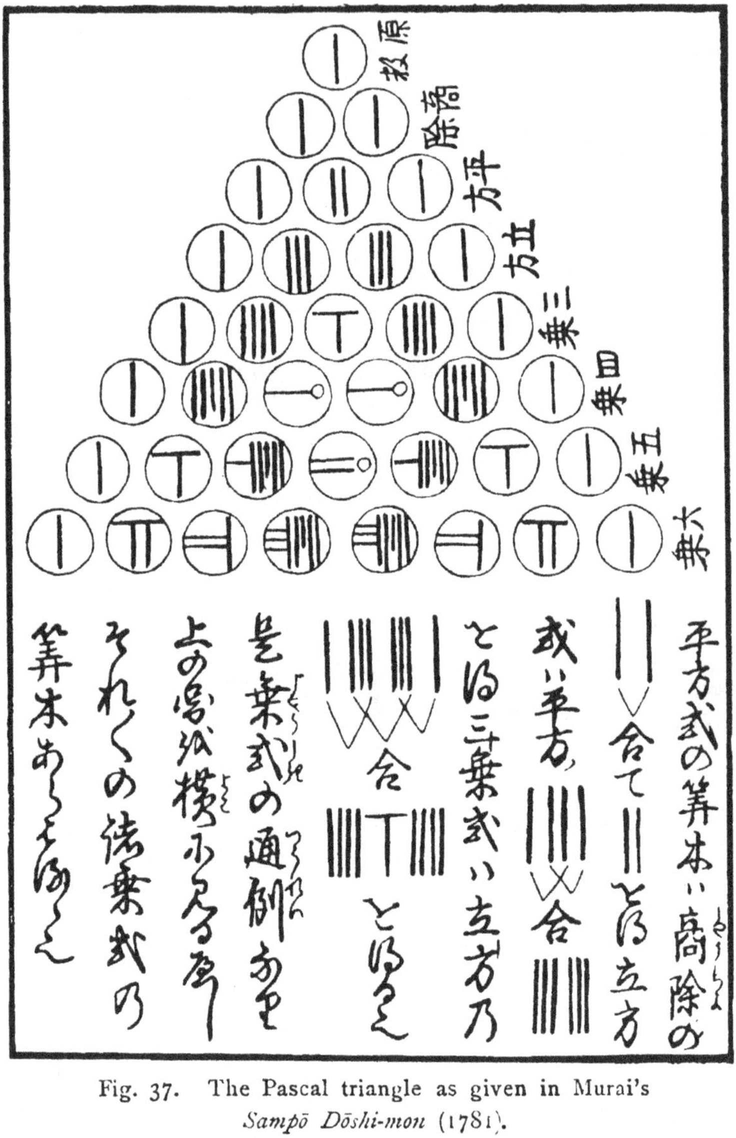

Fig. 37. The Pascal triangle as given in Murai's *Sampō Dōshi-mon* (1781).

he makes no mention of this fact. Nevertheless he assisted in preparing the way for modern mathematics by discouraging the use of the *sangi*, which were already beginning to be looked upon as unwieldy by the best algebraists of his time.

Murai also wrote a *Sampō Dōshi-mon*, or Arithmetic for the Young (see Figs. 36—38), which was intended as a sequel

Fig. 38. From Murai's *Sampō Dōshi-mon* (1781).

to the *Kanja Otogi Zōshi* of Nakane Genjun. The work appeared in 1781, and contains numerous interesting pictures of primitive work in mensuration (Fig. 36), and the Pascal triangle (Fig. 37). It is also noteworthy because of its treatment of circulating decimals. The problem as to the number of figures in the recurring period of a unit fraction was first mentioned in Japan by Nakane in his *Kantō Sampō* (1738) and solutions of an unsatisfactory nature appeared in Ikebe's *Kaishō Sampō* (1743) and in Yamamoto's *Sanzui* (1745). Nakane's writings upon the problem were no longer extant, so that Murai had practically the field before him untouched, although he really did little with it. His theory is brief, for he first divides 9 by 2, 3, ... 9, getting the figures 45, 3, 225, 18, 15, x (not divisible), 1125, 1,—without reference to the decimal points. He then concludes that if unity is divided by 45, 3, 225, ..., the result will have one-figure repetends. Similarly he divides 99 by 2, 3, ... 9, getting the figures 495, 33, 2475, 189, ..., and then divides unity by these results, getting two-figure repetends.

In his explanation of the use of the *soroban* Murai gives certain devices that his predecessors had not in general used. For example, in extracting the square root he divides half of the remainder by the part of the root already found, which he evidently thought to be a little easier on the *soroban* than to divide by twice this root. In treating of cube root he proceeds in an analogous fashion, dividing a third of the remainder twice by the part of the root already found. We have said that these devices had not been used in general before Murai, but they had already been given by at least one writer, Yamamoto Hifumi, in his *Hayazan Tebikigusa*[1] in 1775.

Contemporary with Nakane Genkei, and a friend of his, was a curious character named Kurushima Yoshita, a native of Bitchū, at one time a retainer of Lord Naitō, and a man of notorious eccentricity and looseness of character. It is related of him that when he had to leave Kyūshū to take up his residence in Yedo, he used all of his mathematical manuscripts to repair his basket trunks for the journey. He must, however, have been a man of mathematical ability,

[1] Literally, Handbook of Rapid Calculations.

for he was the friend not only of Nakane but also of Matsunaga, and he had at least one pupil of considerable attainments, Yamaji Shujū. He died in 1757. Among the fragments of knowledge that have been transmitted concerning him is a formula for the radius r, of a regular n-gon of side s, expressed in an infinite series.[1]

Kurushima also knew something of continued fractions, since in Ajima's *Fukyū Sampō*[2] and other works it is shown how he expressed a square root in this manner, with the method of finding the successive convergents. This seems to have been an invention made by him in 1726.[3] It is repeated in a work written in 1748 by Hasu Shigeru, a pupil of one Horiye who had learned from Takebe. In the preface Horiye says that the method is one of the most noteworthy of his time.[4]

Kurushima was also interested in magic squares, and his method of constructing one with an odd number of cells is worth repeating.[5]

The plan may briefly be described as follows:

Let n be the number of cells in one side. Arrange the

---

1 Endō, Book II, p. 112; Kawakita in the *Honchō Sūgaku Kōenshū*, p. 6. On the life of Kurushima there is a manuscript (Japanese) entitled *Tea-table Stories told by Yamaji*. This formula was first published in Aida Ammei's *Sampō Kokon Tsūran* (General View of Mathematical Works ancient and modern), 1795, Book VI. It appears again in Chiba's *Sampō Shinsho* (New Treatise on Mathematical Methods). See Fukuda, *Sampō Tamatebako*. Book II, p. 33; Endō, Book III, p. 33. Kurushima also wrote the *Kyūshi Kohai Sō* (Incomplete Fragments on the arc of a circle) in which he treated of the minimum ratio of an arc to its altitude. It exists only in manuscript. In it is also some work in magic cubes.

2 In manuscript, compiled by Kusaka.

3 Possibly Takebe was the first Japanese to employ continued fractions, in his *Fukyū Tetsujutsu* (1722). See also the *Taisei Sankyō*, where they are found. But their application to square root begins, in Japan, with Kurushima. C. Kawakita relates in the *Sūgaku Hōchi* that this was done in the first month of 1726.

4 Horiye's preface to Hasu's *Heihō Reiyaku Genkai*, 1748, in manuscript. See also Endō, Book II, p. 105.

5 It is given in his manuscript *Kyūshi Ikō* (Posthumous Writings of Kurushima), Book I.

numbers 1, n^2, n, and $k = n^2 + 1 - n$ as in the figure. Then take $\frac{1}{2}(n^2 + 1)$ as the central number, and from this, along

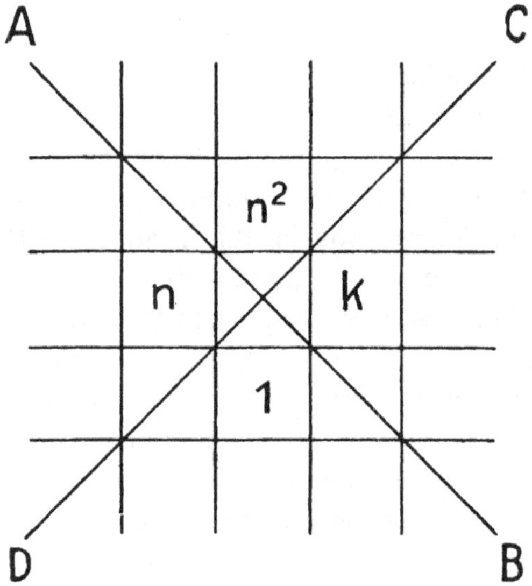

CD, arrange a series decreasing towards *C* and increasing towards *D* by the constant difference n. Next fill the cells along the oblique lines through n and n^2, and through 1 and k, according to the same law. Now fill the cells along *AB* and the two parallels through n and 1, and through n^2 and k, by a series decreasing towards *A* and increasing towards *B* by the constant difference 1. The rest of the rule will be apparent by examining the following square:

22	47	16	41	10	35	4
5	23	48	17	42	11	29
30	6	24	49	18	36	12
13	31	7	25	43	19	37
38	14	32	1	26	44	20
21	39	8	33	2	27	45
46	15	40	9	34	3	28

It is also worthy of note that Kurushima discussed[1] the problem of finding the maximum value of the quotient of the altitude of a circular segment by its arc. In this there arises the equation

$$4 - x^2 + \frac{x^4}{3.6} - \frac{x^6}{3.5 \times 6.8} + \frac{x^8}{3.5.7 \times 6.8.10} - \frac{x^{10}}{3.5.7.9 \times 6.8.10.12} + \cdots = 0.$$

He speaks of this as an "unlimited equation", and after a complicated solution he reaches the result,

$$x = 5.434131504304.$$

Mention should also be made of a value of π^2 given by Kurushima, $\frac{98548}{9985}$; but his method of obtaining it is not known.[2]

In the first half of the eighteenth century there lived in Ōsaka one Takuma Genzayemon, concerning whose life and early training we know practically nothing. Some have said that he learned mathematics in the school of Miyagi, but all that is definitely known is that he established a school in Ōsaka. He is of interest because of his work upon the value of π, a problem that he attacks in the Dutch manner of a century earlier. He seems to have been the only mathematician in Japan who used for this purpose the circumscribed regular polygon as well as the inscribed one of a large number of sides. He bases his conclusions upon the perimeters of polygons of 17,592,186,044,416 sides which he stated to be

3.14159 26535 89793 23846 26433 6658,
3.14159 26535 89793 23846 26434 67.

He takes the average of these numbers, and thus finds the value correct to twenty-five figures. It is related that this was looked upon as one of the most precious secrets of his

---

[1] In his manuscript entitled *Kyūshi Kohai-sō*.

[2] Endō, Book II, p. 127. It is found in manuscript in the posthumous writings of Kurushima.

school.[1] The most distinguished of Takuma's followers was Matsuoka Nōichi (or Yoshikadsu), who published a very usable textbook in 1808, the *Sampō Keiko Taizen.*[2]

Mention has already been made of Matsunaga Ryōhitsu,[3] but his work is such as to merit further notice. One of his most important treatises is embodied in a manuscript called the *Sampō Shūsei*,[4] consisting of nine books of which the first five are devoted to indeterminate analysis as applied to questions of geometry. He considers, for example, the Pythagorean triangle of sides a, b, and hypotenuse c, and lets

$$a = 2m + 1, \quad c - b = 2n,$$

whence

$$c + b = \frac{c^2 - b^2}{c - b} = \frac{a^2}{c - b} = \frac{(2m+1)^2}{2n},$$

whence b and c assume the form

$$\frac{1}{2}\left[(c+b) \mp (c-b)\right] = \frac{(2m+1)^2}{4n} \mp n.$$

Hence the three sides may be represented by

$$4n\,(2m+1),\ (2m+1)^2 - 4n^2,\ (2m+1)^2 + 4n^2.$$

He also attacks the problem by letting the perpendicular p from the vertex of the right angle cut the hypotenuse into the segments c' and c''. He then gets

$$b^2 - a^2 = (c''^2 + p^2) - (c'^2 + p^2)$$
$$= (c'' + c')(c'' - c') = c\,(c'' - c'),$$
$$2ab = c.2p,$$

and $$a^2 + b^2 = c^2.$$

Then since $p^2 = c'c''$, we have

$$(c'' - c')^2 + (2p)^2 = c^2,$$

---

[1] ENDŌ, T., *On the development of the mensuration of the circle in Japan* (in Japanese), *Rigakkai*, Book III, no. 4.

[2] Literally, A Complete Treatise of mathematical instruction.

[3] See page 158. The name also appears as Matsunaga Yoshisuke.

[4] Literally, A Collected Treatise on mathematical methods. It is undated. His Hōyen Sankyō is dated 1739 in one of the prefaces and 1738 in another.

whence the sides of a right triangle may be represented by

$$b^2 - a^2,\ 2ab,\ \text{and}\ a^2 + b^2.$$

Matsunaga was, like most of his contemporary geometers, interested in the radius of the regular polygon of n sides, each side being equal to s. His formula,

$$r^2 = \frac{62370n^4 + 107480n^2 + 83577}{2462268n^2 - 3857400} \cdot s^2,$$

is claimed to give the radius correct to six figures.[1] A more complicated formula, requiring the extraction of a seventh root, is given in Irino Yōshō's *Kakusō Sampō* (1743), but it is no more accurate.

Still another formula of this nature is given by Matsunaga's pupil Yamaji Shujū (1704—1772)[2],

$$\begin{aligned} r^2 = (&151762163981 0n^8 + 1004974720807n^6 \\ &+ 166374503856n^4)\, s^2 \div (5991320086184 1n^6 \\ &- 157432047580066n^4 + 135529756473206n^2 \\ &- 35692069491815). \end{aligned}$$

Such efforts, however, are interesting chiefly for the same reason as the Japanese ivory carving of spheres within spheres, —examples of infinite painstaking. Yamaji was a native of the province of Bitchū, and later he became a *samurai* of the shogunate, serving as assistant in the astronomical department. He first studied under Nakane, and upon Nakane's leaving Yedo for Kyōto he came under the latter's friend Kurushima. When Kurushima moved to Kyūshū, Yamaji became a pupil of Matsunaga. He was thus, as he relates in his *Tea-table Stories*, privileged to know the mathematical secrets of three of the best teachers of Japan. While he was not himself a great contributor to the science, he proved to be a great teacher, so that when he died not a few sucessful mathe-

---

[1] The reader may consider it for $n = 4$, $s = \sqrt{2}$, $r = \frac{1}{2}$. It is also given in Arima's *Hōyen Kikō* (1766), but credit is there given to Matsunaga. See also ENDŌ, Book II, p. 109.

[2] ARIMA, *Hōyen Kikō*; ENDŌ, Book II, p. 108.

maticians were counted among his pupils, including Lord Arima, Fujita, and Ajima. It is possible that the *Kenkon no Maki* was written by him, and also the *Kohai no Ri* and other manuscripts on the *yenri*, but the *Gyokuseki Shin-jutsu*[1] is the only work of importance that is certainly his. In this is given a treatment of the volume of the sphere by a kind of integration much like that to be found in the anonymous[2] *Kigenkai.*

Of Yamaji's pupils the first above mentioned was Arima[3] Raidō (1714—1783), Lord of Kurume in Kyūshū. It was he, it will be recalled, who first published the *tenzan* algebra that had been kept a secret in the Seki school since the days of the founder. His *Shūki Sampō* in five books was published in 1769 under the *nom de plume* of Toyota Bunkei, possibly the name of one of his vassals. The work must certainly have been Arima's, however, since only a man in his position would have dared to reveal the Seki secret. In this treatise Arima sets forth and solves one hundred fifty problems, thus being the first noted writer to break from the old custom of solving the problems of his predecessors and setting others for those who were to follow. His questions related to indeterminate analysis, the various roots of an equation, the algebraic treatment of geometric propositions, binomial series, maxima and minima, and the mensuration of geometric figures, including problems relating to tangent spheres (Fig. 39). The curious Japanese manner of representing a sphere by a circle with a lune on one side is seen in Fig. 39. In this work appears a fractional value of π,

$$\pi = \frac{42822\ 45933\ 49304}{13630\ 81215\ 70117},$$

that is correct to twenty-nine decimal places. Arima also wrote several other works, including the *Hōyen Kikō* (1766)[4] and the *Shōsa San-yō* (1764), but none of these was published.

---

1 Literally, The Exact Method for calculating the volume of a sphere.

2 Or *Yenri Kenkon Sho.*

3 Not Akima, his ancestor, as is sometimes stated.

4 In this is also given the value of π mentioned above, and the powers of π from π^2 to π^{22} for the first thirty-two figures.

Among the vassals of Lord Arima was a certain Honda Teiken (1734—1807), who was born in the province of Musashi. He is known in mathematics by another name, Fujita Sadasuke, which he assumed when he came to manhood, a name that acquired considerable renown in the latter half of the eighteenth century. As a youth he studied under Yamaji, and even when he was only nineteen years of age he became, on

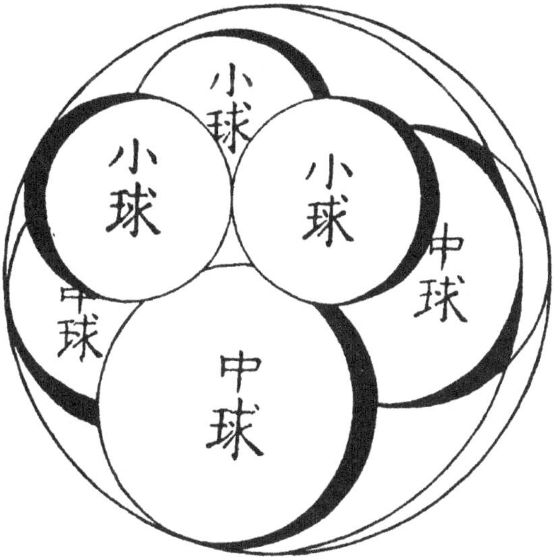

Fig. 39. From Arima's *Shūki Sampō* (1769).

Yamji's recommendation, assistant to the astronomical department of the shogunate. For five years he labored acceptably in this work, but finally was compelled to resign on account of trouble with his eyes. Arima now extended to him a cordial invitation to accompany him to Yedo, whither he went for service every second year, and to act as teacher of arithmetic.[1] Here he published his *Seiyō Sampō* (1779), a work in three books, consisting of a well arranged and carefully selected set of problems in the *tenzan* algebra. This book was so clearly written as to serve as a guide for teachers for a long time after its publication. In Fig. 40 is shown one of his problems

---

[1] Kawakita, in the *Honchō Sūgaku Kōenshū*, 1908, p. 8.

relating to tangent spheres in a cone. Fujita also published several other works, including the *Kaisei Tengen Shinan* (1792),[1] and wrote numerous manuscripts that were eagerly sought by the mathematicians of his time, although of no great merit on the ground of originality. He died in 1807 at the age of seventy-two years, respected as one of the leading mathematicians of his day, although he did not merit any such standing in spite of his undoubted excellence as a teacher.

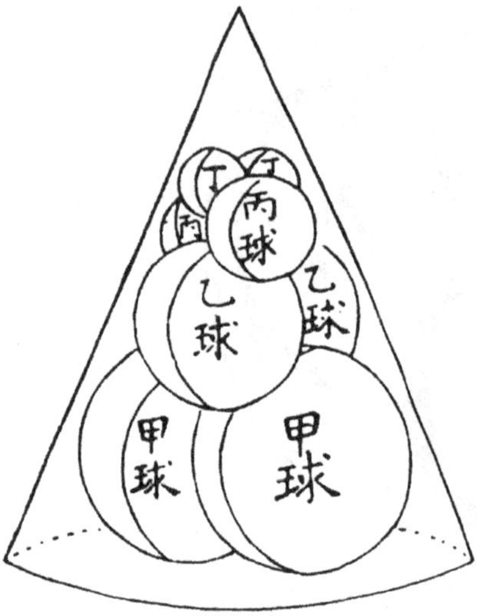

Fig. 40. From Fujita Sadasuke's *Seiyō Sampō* (1779).

Fujita's son Fujita Kagen (1765—1821) was also a mathematician of some prominence. He published in 1790 his *Shimpeki Sampō* (Mathematical Problems suspended before the Temple),[2] and in 1806 a sequel, the *Zoku Shimpeki Sampō*. The significance of the name is seen in the fact that the work contains a collection of problems that had been hung before various temples by certain mathematical devotees between 1767 and the time when Fujita wrote, together with rules for their solution. This strange custom of hanging problems before the temples originated in the seventeenth century, and continued for more than two hundred years. It may have arisen from a desire for the praise or approval of the gods, or from the fact that this was a convenient means of publishing a discovery, or from the wish to challenge others to solve a problem, as European students in the Middle Ages would post a thesis on the door of church. A few of these

---

1 We follow Endō. Hayashi gives 1793.

2 There was a second edition in 1796, with some additions.

problems are here translated[1] as specimens of the work of Japanese mathematicians at the close of the eighteenth century.

"There is a circle in which a triangle and three circles, *A*, *B*, *C*, are inscribed in the manner shown in the figure.

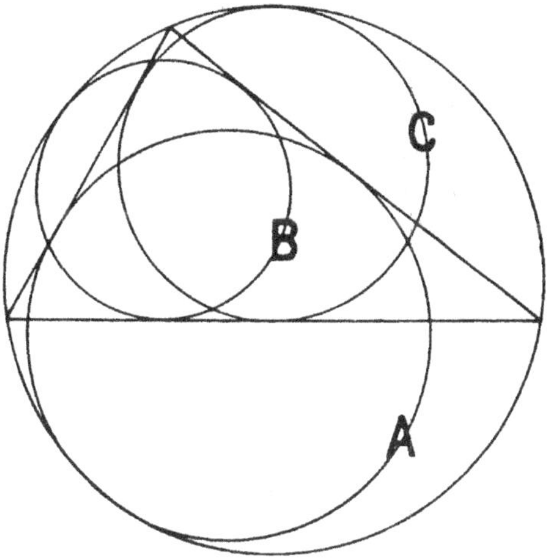

Given the diameters of the three inscribed circles, required the diameter of the circumscribed circle." The rule given may be abbreviated as follows:

Let the respective diameters be x, y, and z, and let $xy = a$. Then from a^2 take $\left[(x-y)z\right]^2$. Divide a by this remainder and call the result b. Then from $(x+y)z$ take a, and divide 0.5 by this remainder and add b, and then multiply by z and by a. The result is the diameter of the circumscribed circle.[2] To this rule is appended, with some note of pride, the words: "Feudal District of Kakegawa in Yenshū Province, third month of 1795, Miyajima Sonobei Keichi, pupil of Fujita Sadasuke of the School of Seki."

Another problem is stated as follows: "Two circles are described, one inscribing and the other circumscribing a quadri-

---

1 From the edition of 1796.

2 That is

$$xyz\left[\frac{xy}{x^2y^2 - [(x-y)z]^2} + \frac{0.5}{(x+y)z - xy}\right].$$

lateral. Given the diameter of the circumscribed circle and the product of the two diagonals, required to find the diameter of the inscribed circle." The problem was solved by Kobayashi Kōshin in 1795, and the relation was established that

$$i\sqrt{c+p}=p,$$

where i = the diameter of the inscribed circle, c = the diameter of the circumscribed circle, and p = the given product.[1]

A third problem is as follows: "There is an ellipse in which five circles are inscribed as here shown. The two axes of

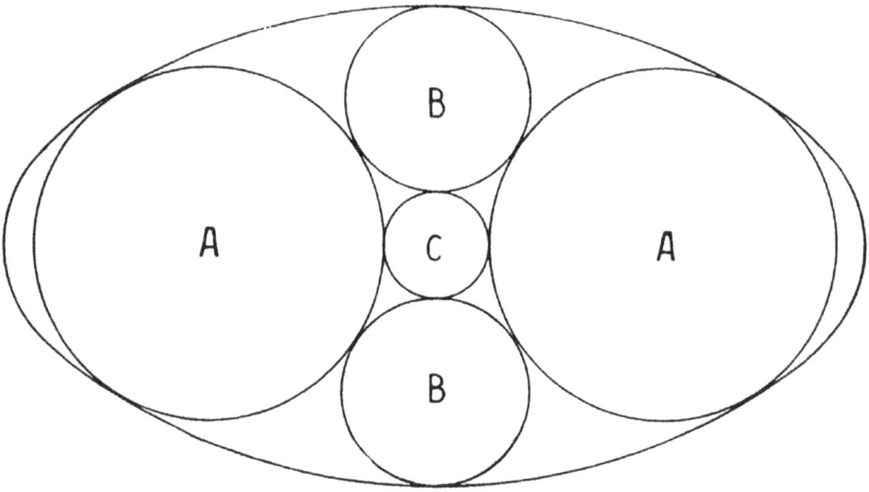

the ellipse being a and b it is required to find the diameter of the circle A." The solution as given by Sano Ankō in 1787 may be expressed as follows:

$$d=b-\frac{2b^3}{\frac{3a^2+b^2}{2}+\sqrt{\left(\frac{3a^2-b^2}{2}\right)^2-a^4}}$$

Another problem of similar nature is shown in Fig. 41, from the *Zoku Shimpeki Sampō* (1806).

A style of problem somewhat similar to one already mentioned in connection with Arima was studied in 1789 by Hata

[1] For the case of a square of side 2 we have $2\sqrt{16}=8$.

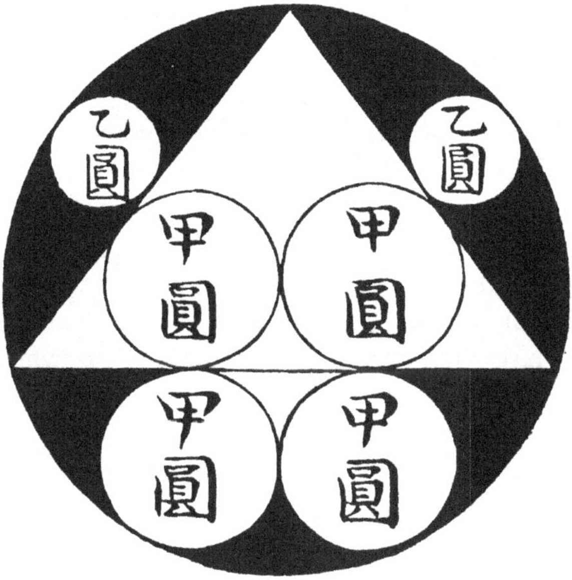

Fig. 41. From the *Zoku Shimpeki Sampō* (1806).

Jūdō, as follows: "There is a sphere in which are inscribed, as in the figure, two spheres *A*, two *B*, and two *C*, touching each

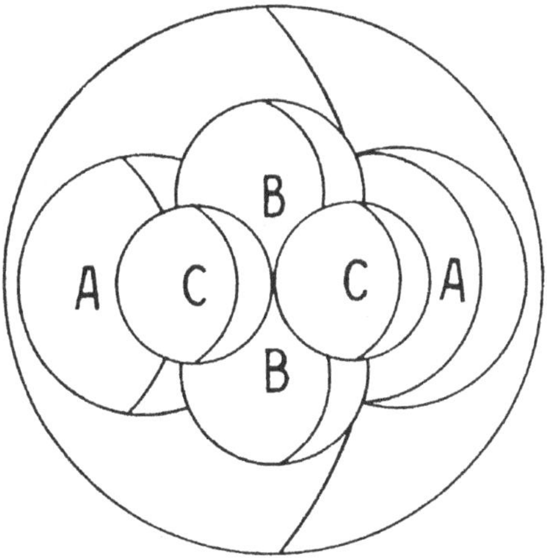

other as shown. Given the diameters of *A* and *C*, required to find the diameter of *B*." The solution given is

$$b = 6a \div \left(1 + \sqrt{\frac{a}{b}}\right).$$

Contemporary with Fujita Sadasuke was Aida Ammei (1747—1817), who was born at Mogami, in north-eastern Japan. Like Seki, Aida early showed his genius for mathematics, and while still young he went to Yedo where he studied under a certain Okazaki, a disciple of the Nakanishi school, and also under Honda Rimei, although he used later to boast that he was a self-made mathematician, and to assume a certain conceit that hardly became the scholar. Nevertheless his ability was such and his manner to his pupils was so kind that he attracted to himself a large following, and his school, to which he gave the boastful title of Superior School, became the most popular that Japan had seen, save only Seki's. Aida wrote, so his pupils say, about a thousand pamphlets on mathematics, although only a relatively small number of his contributions are now extant. He died in 1817 at the age of seventy years.[1]

One of Aida's works, the *Tōsei Jinkōki* (1784) deserves special mention for its educational significance. In this he discarded the inherited problems to a large extent and substituted for them genuine applications to daily life. The result was a great awakening of interest in the teaching of mathematics, and the work itself was very successful.

Soon after the publication of this work there arose an unfortunate controversy between Aida and Fujita Sadasuke, at that time head of the Seki School. The story goes[2] that Aida had at one time asked to be admitted to this school, but that Fujita in an imperious fashion had told him that first he must make haste to correct an error in his solution of a problem that he had hung in the Shintō shrine on Atago hill in Shiba, Tōkyō. Aida promptly declined to change his solution and thus cut himself off from the advantages of study in the Seki school. While Aida admits having visited Fujita he says that he did so only to test the latter's ability, not for the purpose of entering the school.

[1] As stated upon his monument. See also C. Kawakita in the *Honchō Sūgaku Kōenshū*, 1908.

[2] This account is digested from the works of various writers who were drawn into the controversy.

As a result of all this unhappy discussion Aida was much embittered against the Seki school, and in particular he set about to attack the *Seiyō Sampō* which Fujita Sadasuke had published in 1779. For this purpose he wrote the *Kaisei Sampō*, or *Improved Seiyō Sampō*, and published it in 1785, criticising severely some thirteen of Fujita's problems, and starting a controversy that did not die for a score of years. Fujita's pupil, Kamiya Kōkichi Teirei, then wrote in the former's defence the *Kaisei Sampō Seiron*, and sent the manuscript to Aida, to which the latter replied in his *Kaisei Sampō Kaisei-ron* which appeared in 1786. Kamiya having been forbidden by Fujita to publish his manuscript, so the story runs, he prepared another essay, the *Hi-kaisei Sampō* which also appeared about the same time, the exact date being a subject of dispute. Of the replies and counter-replies it is not necessary to speak at length, since for our purposes it suffices to record this Newton-Leibnitz quarrel in miniature.[1] It was in one sense what is called in English a "tempest in a tea-pot"; but in another sense it was more than that, for it was a protest against the claims of the Seki school, of the individual against the strongly entrenched guild, of genius against authority, of struggling

[1] For purposes of reference the following books on the controversy are mentioned: Fujita wrote a reply to Aida in 1786, which was never printed. Aida wrote the *Kaiwaku Sampō* in 1788, replying to the *Hi-kaisei Sampō*. Fujita wrote a rejoinder, the *Hi-kaiwaku Sampō*, but it was never printed. Kamiya published the *Kaiwaku Bengo* in 1789, replying to Aida. In 1792 Aida wrote the *Shimpeki Shinjutsu* in which he criticised the *Shimpeki Sampō* of Fujita's son, and also wrote the *Kaisei Sampō Jensho* in which he criticised Fujita's *Seiyō Sampō*, but neither of these was printed. In 1795 he wrote his *Sampō Kakujo*, an abusive reply to Kamiya, but in the same year he wrote the *Sampō Kokon Tsūran* (General view of mathematical works, ancient and modern) in which he has something good to say of him. In 1799 Kamiya wrote an abusive reply to Aida, the *Hatsuran Sampō*. The last of the published works by the contestants was Aida's *Hi-Hatsuran Sampō* of 1801, although the controversy still went on in unpublished manuscripts. The manuscripts include Kamiya's *Fukusei Sampō* (1803) and Aida's *Sampō Senri Dokkō* (1804). Mention should also be made of the *Sampō Tenshō hō Shinan* (1811) written by Aida, of which only the first part (5 books) was printed.

youth against vested interests; it was the cry of the insurgent who would not be downed by the abuse of a Kamiya who championed the cause of a decadent monopoly of mathematical learning and teaching. It was this that inspired Aida to act, and of the dignity of his action these words, from a preface to one of his works, will bear witness: "The *Seiyō Sampō*[1] treats of subjects not previously worked out, and certain of its methods have never been surpassed. The author's skill in mathematics may safely be described as unequalled in all the Empire. Upon this work the student may in general rely, although it is not wholly free from faults. Since it would be a cause of regret, however, if posterity should be led into error through these faults, as would be the natural influence of so great a master as Fujita, I have taken the trouble to compose a work which I now venture to offer to the world as a guide." Such words and others in recognition of Fujita's merits did not warrant the abuse that Kamiya heaped upon Aida, and the impression left upon the reader of a century later is that of a staunch champion of liberty of thought, combatted by the unprovoked insults and unjust scorn of vested interests. Fujita seems to have solved his problems correctly but to have expressed his work in cumbersome notation,[2] while Aida stood for simplicity of expression. Neither was in general right in attacking the solutions of the other, and in the heat of controversy each was led to statements that were incorrect. The whole struggle is a rather sad commentary on the state of mathematics in the waning days of the Seki school, when the trivial was magnified and the large questions of mathematics were forced into the background.

Aida was an indefatigable worker, practically his whole life having been spent in study. As a result he left hundreds of manuscripts, most of which suffered the fate of so many

---

[1] Fujita's work of 1779.

[2] As compared with that of Aida, although an improvement upon that of his predecessors.

thousands of books in Japan, the fate of destruction by fire.[1] Of the contents of the *Sampō Kokon Tsūran* (1795) already mentioned, only a brief note need be given. In Book VI Aida gives the value of $\frac{\pi}{2}$ as follows:

$$\frac{\pi}{2} = 1 + \frac{1!}{3} + \frac{2!}{3\cdot5} + \frac{3!}{3\cdot5\cdot7} + \frac{4!}{3\cdot5\cdot7\cdot9} + \cdots.$$

He gives a series for the length of an arc x in terms of the chord c and height h thus:

$$x = c\left(1 + \frac{2}{3}m + \frac{2.4}{3\cdot5}m + \frac{2.4.6}{3\cdot5\cdot7}m + \cdots\right),$$

where $$m = \frac{h^2}{\left(\frac{1}{2}c\right)^2 + d^2}$$

and d is the diameter of the circle. In the same work he gives a formula for the area of a circular segment of one base:

$$a = \frac{hc}{2}\left(1 + \frac{2}{3}m + \frac{2.4}{3\cdot5}m + \frac{2.4.6}{3\cdot5\cdot7}\cdots\right).$$

Aida also gave a solution of a problem found in Ajima's *Fukyū Sampō*, as follows: The side of an equilateral triangle is given as an integer n. It is required to draw the lines s_1, s_2, . . ., parallel to one side, such that the p's, q's and s's as shown in the figure shall all have integral values.

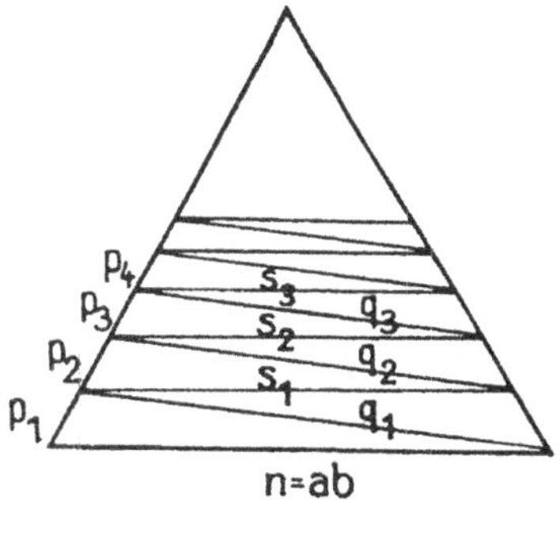

Ajima had already solved this before Aida tried it, and this is, in substance, his solution: Decompose n into two factors, a and b, which are either both odd or both even. If this cannot be done a solution is impossible. The rules are now, as expressed in formulas, as follows:

[1] KAWAKITA's article in the *Honchō Sūgaku Kōenshu*, p. 13.

$$p_1 = k^2 - a^2, \qquad q_1 = (k-a)^2 - ka,$$
$$p_2 = p_1 - D, \qquad p_3 = s_2 - D, \ldots$$
$$s_2 = n - p_1, \qquad s_3 = s_2 - p_2, \ldots$$
$$q_2 = q_1 + M - p_1, \qquad q_3 = q_2 + M - p_2, \ldots$$

where $k = \frac{1}{2}(a+b)$, $D = \frac{1}{2}(b-a)^2$, $M = \frac{1}{2}D$.

When $p_1 > \frac{1}{2}n$ it may be taken at once for s_2 and $n - s_2$ for p_1.[1]

Aida objects to the length of such a rule, and he proposes to solve the problem thus:

Let $n = ab$, where $a < b$.

Then let $\frac{1}{2}(b-a) = D$.

Then $(a-D)(b-D) = s_2$,

$(a-2D)(b-2D) = s_3, \ldots .$

Also let $s_{r-1} - s_r = p_r$,

and we have

$$\left(\frac{a-rD}{2}\right)^2 + 3\left(\frac{b-rD}{2}\right)^2 = q_r.$$

Aida also did some work in indeterminate equations[2] and was the first to take up the permutation of magic squares.[3]

---

[1] Ajima does not tell what to do for q_1 if $(k-a)^2 < ka$.

[2] As in solving $z^2 = x_1^2 + x_2^2 + x_3^2 + x_4^2 + x_5^2$. See the article by C. Hitomi in the *Journal of the Tōkyō Physics School.* From Aida's manuscript *Sampō Seisū-jutsu* (On the method of solutions in integers), we also take the following types:

$$1^2 x_1^2 + 2^2 x_2^2 + 3^2 x_3^2 + \cdots + 10^2 x^2_{10} = y^2$$

and

$$1 x_1^2 + 2 x_2^2 + 3 x_3^2 + \cdots + 10 x^1_{10} = y^2.$$

This manuscript was probably written not earlier than 1807.

[3] Upon the authority of K. Kano, to whom we are indebted for the statement.

He also gives an ingenious method for expanding a binomial, or rather for writing down the coefficients in the expansion of $(a+b)^{\frac{1}{n}}$, which expresses roots in series.

One of the most interesting of Aida's solutions is that of the problem to find the radius r of a regular n-gon of side s.[1] He says that of the infinite series representing $\frac{s}{r}$ the successive terms are

$$\frac{6}{n},\quad \frac{\left[1^2-\left(\frac{6}{n}\right)^2\right]\cdot\frac{6}{n}}{4.6},\quad \frac{\left[1^2-\left(\frac{6}{n}\right)^2\right]\left[3^2-\left(\frac{6}{n}\right)^2\right]\cdot\frac{6}{n}}{4.6.8.10},$$
$$\frac{\left[1^2-\left(\frac{6}{n}\right)^2\right]\left[3^2-\left(\frac{6}{n}\right)^2\right]\left[5^2-\left(\frac{6}{n}\right)^2\right]\cdot\frac{6}{n}}{4.6.8.10.12.14},\ \ldots.$$

If we put m for $\frac{6}{n}$, and x for $\frac{1}{2}$, the series becomes

$$\frac{\sin(m \text{ arc } \sin x)}{x} = \frac{m}{1!} - \frac{m(m^2-1^2)}{3!}x^2 + \frac{m(m^2-1^2)(m^2-3^2)}{5!}x^4 \mp \ldots,$$

a series that has been attributed both to Newton and to Euler. We therefore have

$$\frac{s}{r} = 2\sin\left(\frac{6}{n}\text{ arc } \sin\frac{1}{2}\right),$$

or

$$\frac{s}{r} = 2\sin\frac{\pi}{n},$$

whence $\sin\frac{\pi}{6}=\frac{1}{2}$. It is generally conceded that Aida knew that the formula had already been given in substance by Kurushima.[2] It also appeared in Matsunaga's *Hōyen Sankyō* of 1739.

From the names considered in this chapter we might characterize the eighteenth century as one of problem-solving, of the extension of a rather ill-defined application of infinite series

---

[1] Hayashi, *History*, part II, p. 13.

[2] See p. 176.

to the mensuration of the circle, of some slight improvement in the various processes, of the rather arrogant supremacy of the Seki school, and of a bitter feud between the independents and the conservatives in the teaching of mathematics. And this is a fair characterization of most of the latter half of the century. There was, however, one redeeming feature, and this is found in the work of Ajima Chokuyen, of whom we shall speak in the next chapter.

CHAPTER X.

Ajima Chokuyen.

In the midst of the unseemly strife that waged between Fujita and Aida in the closing years of the eighteenth century there dwelt in peaceful seclusion in Yedo a mathematician who surpassed both of these contestants, and who did much to redeem the scientific reputation of the Japanese of his period. A man of rare modesty, content with little, taking delight in the simple life of a scholar rather than in the attractions of office or society, almost unknown in the midst of the turmoil of the scholastic strife of his day, Ajima Manzō Chokuyen[1] was nevertheless a rare genius, doing more for mathematics than any of his contemporaries.

He was born in Yedo in 1739, and as a *samurai* he served there under the Lord of Shinjō, whose estates were in the north-eastern districts. He was initiated into the secrets of mathematics by one Iriye Ōchū[2], who had studied in the school of Nakanishi. He afterwards became a pupil of Yamaji Shujū, and at this time he came to know Fujita Sadasuke with whom he formed a close friendship but with whose controversy with Aida he never concerned himself. And so he received a training that enabled him to surpass all his fellows in solving the array of problems that had accumulated during the century, including all those which had long been looked upon as wholly insoluble. Such a type of mind rarely extends the boundaries of mathematical discovery, but occasionally an individual is

[1] See also HARZER, P, loc. cit., p. 34 of the Kiel reprint of 1905.

[2] Also given as Irie Masatada.

found with this kind of genius who is at least able to help in improving science by his genuine sympathy if not by his imagination. Such a man was Ajima. His interests extended from *tenzan* algebra to the Diophantine analysis, and from simple trigonometry to a new phase of the *yenri* which had occupied so much attention throughout the century. Possessed of the genius of simplicity, he clothed in more intelligible form the abstract work of his predecessors, even if he made no noteworthy discovery for himself. Although his retiring nature would not allow him to publish his works, he left many manuscripts of which the more important may well occupy our attention. He died in 1798 at the age of fifty-nine years,[1] honored by his fellows as a *Meijin*[2] (genius, or person dexterous in his art) in the field in which he labored.

In the *Kan-yen Muyūki*[3] (1782) he gives a solution in integers of the problem of n tangent circles described within a given circle, and similarly for an array of circles tangent to one another and to the given circle externally. The problem is one of those in indeterminate analysis to which the Japanese scholars paid much attention. Another indeterminate equation considered by him is the following:

$$x_1^2 + x_2^2 + x_3^2 + x_4^2 + x_5^2 = y^2.$$

This appears in a manuscript entitled *Beki-wa Kaihō Mu-yūki Seisū-jutsu* (Integral solutions for the square root of the sum of squares) and dated 1791.

Another work of his was the *Sampō Kōsō*,[4] in which the famous Malfatti problem appears, to inscribe three circles in a triangle, each tangent to the other two. Ajima does not, however, consider the geometric construction, preferring to attack the question from the standpoint of algebra, after the usual manner of the Japanese scholars. The problem first

1 C. Kawakita, in his article in the *Honchō Sūgaku Kōenshū* says that he is sometimes thought to have died in 1800, but the date given by us is from the records of the Buddhist temple where he is buried.

2 The term may be compared to *pandit* in India.

3 Literally, Integral solutions of circles touching a circle.

4 Literally, A draft of a mathematical problem.

appears in Japan, so far as now known, in the *Sampō Gakkai*[1] published by Ban Seiyei of Ōsaka in 1781, the solution being much more complicated than that given subsequently by Ajima.[2]

The *Senjō Ruiyen-jutsu*[3] and the *Yennai Yō-ruiyen-jutsu*[4] are two works upon groups of circles tangent to a straight line and a circle, or to two circles. In the *Renjutsu Henkan* (1784)[5] he treats the subject still more generally, considering the straight line as a limiting case of a circumference.

The *Jūji-kan Shinjutsu*,[6] a manuscript of 1794, considers the question of an anchor-ring cut by two cylinders, a problem first studied in Japan by Seki, and later by Arima in his *Shūki Sampō* (1769), where infinitesimal analysis seems to have been applied to it for the first time in this country. One of the most famous problems solved by Ajima is that known as the Gion Temple Problem, and treated by him in his *Gion Sandai no Kai.*[7] The problem is as follows: "There is a segment of a circle, and in this there are inscribed, on opposite sides of the altitude, a circle and a square. Given the sum of the chord, the altitude, the diameter of the inscribed circle, and a

---

[1] Literally, Sea of learning for mathematical methods.

[2] Endō, Book III, p. 187. For the history of the problem in the West see A. Wittstein, *Geschichte des Malfatti'schen Problems,* München, 1817, Diss.; M. Baker in the *Bulletin of the Philosophical Society of Washington,* Vol. II, p. 113; *Intorno alla vita ed agli scritti di Gianfranco Malfatti,* in the Boncompagni *Bulletino,* tomo IX, p. 361. For the isosceles triangle the problem appears in the *Opera* of Jakob Bernoulli, Geneva, 1744, *Problema geometrica,* lemma II, tomus I, p. 303. It was first published by Malfatti (1731—1807) in the *Memorie di Matematica e di Fisica,* Modena, 1803, tomo X, p. 235, five years after Ajima died.

[3] Literally, On Circles described successively on a line. It appeared in 1784, and a sequel 1791.

[4] Literally, On Circles described successively within a circle.

[5] Literally, The Adapting of a general plan to special cases.

[6] Literally, Exact method for the cross-ring.

[7] Literally, The Analysis of the Gion Temple problem. The manuscript is dated the 24th day of the 6th month, 1773, although Endō (Book III, p. 8) gives 1774 as the year.

side of the square, and also given the sum of the quotients of the altitude by the chord, of the diameter of the circle by

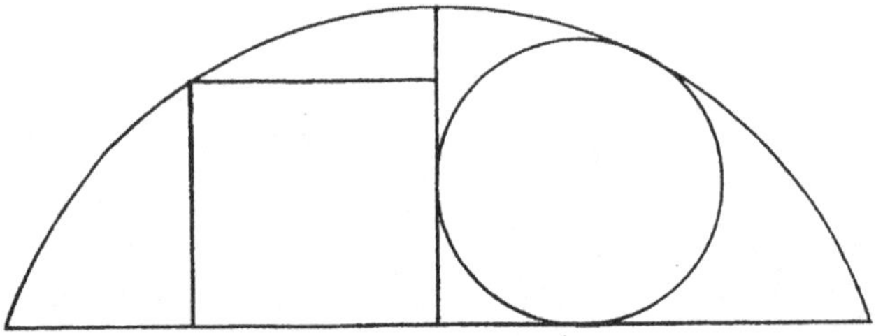

the altitude, and of the side of the square by the diameter of the circle, it is required to find the various quantities mentioned."

The problem derives its name from the fact that it was, with its solution, first hung before the Gion Temple in Kyōto by Tsuda Yenkyū, a pupil of Nishimura Yenri's[1], the solution depending upon an equation of the 1024th degree in terms of the chord. The solution was afterward simplified by one Nakata so as to depend upon an equation of the 46th degree. Ajima attacked the problem in the year 1774, and brought it down to the solution of an equation of the 10th degree. This is not only a striking proof of Ajima's powers of simplification, but it is also evidence of the improvement constantly going on in the details of Japanese mathematics in the eighteenth century.

Ajima considers in his *Fujin Isshū* (Periods of decimal fractions) the problem of finding the number of figures contained in the repetend of a circulating decimal when unity is divided by a given prime number. Although he states that the problem is so difficult as to admit of no general formula, he shows great skill in the treatment of special cases. To assist him he had the work of at least two predecessors, for Nakane Genjun had studied the problem for special cases in his *Kantō Sampō* of 1738, and in the *Nisei Hyōsen* Ban Seiyei of Ōsaka had given the result for a special case, but without

1 Whose *Tengaku Shiyō* (Astronomy extract) was published in 1776.

the solution. Ajima was, however, the first Japanese scholar to consider it in a general way.

He first gives a list of numbers from which, considered as divisors of unity, there arise periods of from 1 to 16 figures, as follows:

1 figure	3		
2 figures	11		
3 figures	37		
4 figures	101		
5 figures	41,	271	
6 figures	7,	13	
7 figures	239,	4649	
8 figures	73	137	
9 figures	333,667		
10 figures	9091		
11 figures	21,649,	513,239	
12 figures	9901		
13 figures	53,	79,	665,371,653
14 figures	909,091		
15 figures	31,	2,906,161	
16 figures	17,	5,882,353.	

As an example of his methods we will consider his treatment of the special fractions $\frac{1}{353}$ and $\frac{1}{103}$. Ajima assumes without explanation that the required numbers are given by one of the possible products of some of the prime factors in

$$353 - 1 = 352 = 2^5 \times 11$$

and

$$103 - 1 = 102 = 2 \times 3 \times 17,$$

respectively. He then says that out of these products it can be found by trial that the respective numbers sought are 32 and 34, but he does not tell how this trial is effected. This was done later by Koide Shūki (1797—1865) and the result appeared in print in the *Sampō Tametebako* (1879), a work by Koide's pupil, Fukuda Sen, who wrote under the *nom de plume* Riken. Koide merely explains Ajima's work, using identically the same numbers.

Neither his explanation nor Ajima's hint is, however, very clear, and each shows both the difficulties met by followers of the *wasan* and their tendency to keep such knowledge from profane minds.

For the expansion of $\sqrt[n]{N}$ Ajima gives two formulas,[1] which may be expressed in modern notation as follows:

$$\sqrt[n]{N} = a + \frac{1}{n} am - \frac{n-1}{2n} D_1 m + \frac{2n-1}{3n} D_2 m - \frac{3n-1}{4n} D_3 m + \cdots$$

$$= a\left[1 + \frac{1}{n} m + \sum_2^\infty (-1)^{i+1} \frac{(n-1)(n-2)\cdots[(i-1)n-1]}{n\cdot 2n\cdot 3n\cdot\cdots in} m^i\right],$$

where $m = \dfrac{N - a^n}{a^n}$.

$$\sqrt[n]{N} = a\left[1 + \sum_1^\infty \frac{1(n+1)(2n+1)\cdots[(i-1)n+1]}{n\cdot 2n\cdot 3n\cdots in} m^i\right],$$

where $m = \dfrac{N - a^n}{N}$. No explanation of the work is given. He also treated of square roots by means of continued fractions, the convergents of which he could obtain.[2]

Ajima also studied the spiral of Archimedes, although not under that name.[3] It had been considered even before Seki's time,[4] and Seki himself gave some attention to it.[5] Lord Arima also discussed it in his *Shūki Sampō* of 1769. It is to Ajima, however, that we are indebted for the only serious treatment up to his time. He divided a sector of a circle by radii into n equal parts, and then divided each of the radii also into n equal parts by arcs of concentric circles. He then joined successive points of intersection, beginning at the center and

---

1 In the *Tetsu-jutsu Kappō* of 1784.

2 HAYASHI, *History*, part II, p. 9, probably refers to his commentary on Kurushima's method.

3 It was called by Japanese scholars *yenkei, yempai,* or *yenwan.*

4 As in Isomura's *Ketsugishō* of 1684.

5 In his *Kai-Kendai no Hō,* and reproduced in the *Taisei Sankyō.*

ending on the outer circle, and said that the limiting form of this broken line for $n = \infty$ was the *yempai*. He then proceeded to find the area between the curve and the original arc by finding the trianguloid areas and summing these for $n = \infty$, obtaining $\frac{1}{3}ar$. In a similar fashion he rectifies the curve, obtaining as a result the series

$$s = r + \frac{a^2}{6r} - \frac{a^4}{40r^3} + \frac{a^6}{112r^5} - \frac{5a^8}{1152r^7} + \frac{7a^{10}}{2816r^9} - \cdots,$$

a result that Shiraishi Chōchū (1822) puts in a form equivalent to

$$s = r\left[1 + \sum_{1}^{\infty}(-1)^{i+1}\frac{1^2\cdot 3^2\cdot 5^2\cdots(2i-3)^2\cdot(2i-1)}{(2i+1)!}m^i\right].$$

Ajima also gives a formula for the square of the length of the curve, and summarizes his work by giving numerical values for $r = 10$, $a = 5$, thus:

$$s = 10.402288144\ldots$$
$$s^2 = 108.2075996685\ldots,$$

from which he concludes that Seki's treatment of the subject was rather crude.

Ajima made a noteworthy change in the *yenri*, in that he took equal divisions of the chord instead of the arc, thus simplifying the work materially.[1] Indeed we may say that in this work Ajima shows the first real approach to a mastery of the idea of the integral calculus that is found in Japan, which approach we may put at about the year 1775. Since this work was so noteworthy we enter upon a more detailed description than is usually required in speaking of the achievements of the eighteenth century.

Ajima proceeds first to find the area of a segment of a circle bounded by two parallel lines and the equal arcs inter-

[1] This appears in his *Kohai-jutsu Kai* (Note on the measurement of an arc of a circle), the date of which is not known. ENDŌ (Book III, p. 1) thinks that it precedes his knowledge of the *yenri* as imparted by his teacher Yamaji.

cepted by them, that is, the area $ABCD$ in the figure. Here we divide the chord c of the arc into n equal parts.[1]

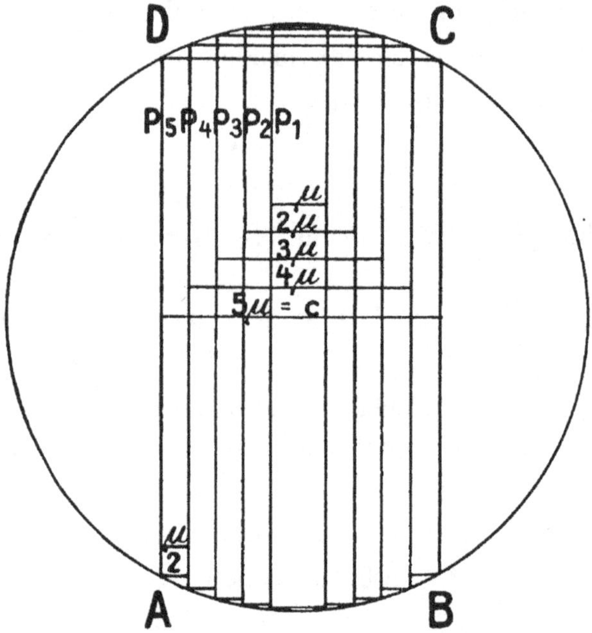

Then from the figure it is apparent that

$$\mu = \frac{c}{n},$$

$$p_r^2 = d^2 - (r\mu)^2,$$

where p_r is the rth parallel from the diameter d.

Ajima now expands p_r, without explaining his process (evidently that of the *tetsujutsu*), and obtains

$$p_r = d\left[1 - \frac{1}{2}\left(\frac{r\mu}{d}\right)^2 - \frac{1}{4\cdot 2}\left(\frac{r\mu}{d}\right)^4 - \frac{3}{4\cdot 3\cdot 2}\left(\frac{r\mu}{d}\right)^6 - \cdots\right]$$

$$= d\left[1 - \frac{1}{2}\left(\frac{r\mu}{d}\right)^2 - \frac{1}{8}\left(\frac{r\mu}{d}\right)^4 - \frac{3}{48}\left(\frac{r\mu}{d}\right)^6 - \frac{15}{384}\left(\frac{r\mu}{d}\right)^8 - \cdots\right].$$

---

[1] In the figure the chord DC is divided into 5 equal parts, each part being designated by μ, so that $5\mu = c$.

Summing for $r = 1, 2, 3 \cdots n$, and multiplying by μ we have the following series:

$$\sum_1^n p_r \mu = d\mu \left[n - \frac{1}{2}\left(\frac{\mu}{d}\right)^2 \Sigma r^2 - \frac{1}{8}\left(\frac{\mu}{d}\right)^4 \Sigma r^4 \right.$$

$$\left. - \frac{3}{48}\left(\frac{\mu}{d}\right)^6 \Sigma r^6 - \frac{15}{384}\left(\frac{\mu}{d}\right)^8 \Sigma r^8 - \cdots \right]$$

$$= d\mu \left[n - \frac{1}{2} \cdot \frac{1}{6}\left(\frac{\mu}{d}\right)^2 (2n^3 + 3n^2 + n) \right.$$

$$- \frac{1}{8} \cdot \frac{1}{30}\left(\frac{\mu}{d}\right)^4 (6n^5 + 15n^4 + 10n^3 - n)$$

$$- \frac{3}{48} \cdot \frac{1}{42}\left(\frac{\mu}{d}\right)^6 (6n^7 + 21n^6 + 21n^5 - 7n^3 + n)$$

$$- \frac{15}{384} \cdot \frac{1}{40}\left(\frac{\mu}{d}\right)^8 (10n^9 + 45n^8 + 60n^7 - 42n^5 + 20n^3 - 3n)$$

$$- \frac{105}{3840} \cdot \frac{1}{66}\left(\frac{\mu}{d}\right)^{10} (6n^{11} + 33n^{10} + 55n^9 - 66n^7 + 66n^5 - 33n^3 + 5n)$$

$$- \frac{945}{46080} \cdot \frac{1}{2730}\left(\frac{\mu}{d}\right)^{12} (210n^{13} + 1365n^{12} + 2720n^{11}$$

$$- 5002n^9 + 8580n^7 - 9009n^5 + 4550n^3 - 691n)$$

$$\left. - \cdots \right].$$

Now substituting for μ its value, $\frac{c}{n}$, and then letting n approach ∞, all terms with n in the denominator approach 0 as a limit, and the limit to which the required area approaches is

$$\text{area} = d \left[c - \frac{1}{6} \cdot \frac{c^3}{d^2} - \frac{1}{40} \cdot \frac{c^5}{d^4} - \frac{3}{336} \cdot \frac{c^7}{d^6} \right.$$

$$- \frac{15}{3456} \cdot \frac{c^9}{d^8} - \frac{105}{43240} \cdot \frac{c^{11}}{d^{10}} - \frac{945}{599040} \cdot \frac{c^{13}}{d^{12}}$$

$$\left. - \cdots \right].$$

From this Ajima easily derives the area of the segment, and from that he gets the length of the arc, as follows:

$$\text{arc} = c + \frac{1^2}{2\cdot 3}\cdot\frac{c^3}{d^2} + \frac{1^2\cdot 3^2}{2\cdot 3\cdot 4\cdot 5}\cdot\frac{c^5}{d^4} + \frac{1^2\cdot 3^2\cdot 5^2}{2\cdot 3\cdot 4\cdot 5\cdot 6\cdot 7}\cdot\frac{c^7}{d^6} + \cdots,$$

from which other formulas may be derived.

Ajima also directed his attention to the problem of finding the volume cut from a cylinder by another cylinder which intersects it at right angles. His result is given by his pupil Kusaka Sei (1764—1839)[1] in his manuscript work, the *Fukyū Sampō* (1799), without explanation, as follows:

$$k^2 d\cdot\frac{\pi}{4}\left[1 - \frac{1}{8}m - \frac{1}{8\cdot 8}m^2 - \frac{1\cdot 5}{8\cdot 8\cdot 16}m^3 - \frac{1\cdot 5\cdot 7}{8\cdot 8\cdot 16\cdot 16}m^4 - \frac{1\cdot 5\cdot 7\cdot 21}{8\cdot 8\cdot 16\cdot 16\cdot 40}m^5 - \cdots\right]$$

where k and d are the diameters of the pierced and piercing cylinders, respectively, and where $m = k^2 \div d^2$.[2] In another work of 1794,[3] however, Ajima gives an analysis of the problem, cutting the solid into elements as in the case of the segment of a circle already described. He then proceeds to the limit as in that case, and thus gives a good illustration of a fairly well developed integral calculus applied to the finding of volumes.[4]

Thus we at last find, in Ajima's work, the calculus established in the native Japanese mathematics, although possibly with considerable European influence. With him the use of the double series again appears, it having already been employed by Matsunaga and Kurushima, and by him the significance of double integration seems first to have been realized. He

---

1 Or Kusaka Makoto.

2 Endō attempts some explanation in his *History*, Book III, p. 25.

3 This is a manuscript of the *Yenchū Sen-kūyen Jutsu* (Evaluation of a cylinder pierced by another).

4 The work as given by Ajima is too extended to be set forth at length, the theory being analogous to that which has already been illustrated.

lacked the simple symbolism of the West, but he had the spirit of the theory, and although his contemporaries failed to realize his genius in this respect, it is now possible to look back upon his work, and to evaluate it properly. As a result it is safe to say that Ajima brought mathematics to a higher plane than any other man in Japan in the eighteenth century, and that had he lived where he could easily have come into touch with contemporary mathematical thought in other parts of the world he might have made discoveries that would have been of far reaching importance in the science.

CHAPTER XI.

The Opening of the Nineteenth Century.

The nineteenth century opened in Japan with one noteworthy undertaking, the great survey of the whole Empire. At the head of this work was Inō Chūkei,[1] a man of high ability in his line, and one whose maps are justly esteemed by all cartographers. Until he was fifty years of age he lived the life of a prosperous farmer. While not himself a contributor to pure mathematics, he came in later life under the influence of the astronomer Takahashi Shiji[2] (1765—1804), and at the solicitation of this scholar he began the work that made him known as the greatest surveyor that Japan ever produced. Takahashi seems to have become acquainted with Western astronomy and spherical trigonometry through his knowledge of the Dutch language. He had also studied astronomy while serving as a young man in the artillery corps at Ōsaka, his teacher having been a private astronomer and diligent student named Asada Gōryū (1732—1799), by profession a physician. This Asada was learned in the Dutch sciences,[3] and is sometimes said to have invented a new ellipsograph.[4] In 1795 he was called to

---

1 Or Inō Tadanori, Inō Tatayoshi, whose life and works are now (1913) being studied by Mr. R. Ōtani.

2 Or Takahashi Shigetoki, Takahashi Yoshitoki, Takahashi Munetoki.

3 As only physicians and interpreters were at this time.

4 A different instrument was invented by Aida Ammei, who left a manuscript work of twenty books upon the ellipse. There is also a manuscript written by Hazama Jūshin in 1828, entitled *Dayen Kigen* (A description of the ellipse) in which it is claimed that the ellipsograph in question was invented by the writer's father, Hazama Jūfū (or Shigetomi) who lived

membership in the Board of Astronomers of the shogunate, an honor which he declined in favor of his pupils Takahashi Shiji and Hazama Jūfū. Takahashi thereupon took up his residence in Yedo, where he died in 1804,[1] five years after Asada had passed away.

Among Asada's younger contemporaries was Furukawa Ujikiyo (1758—1820), who founded a school which he called the *Shisei Sanka Ryū.*[2] He was a shogunate *samurai* of high rank, holding the office of financial superintendent, and although a prolific writer he contributed little of importance to mathematics.[3] Nevertheless his school flourished, although it was one of nineteen[4] at that time contending for mastery in Japan,

---

from 1756 to 1816, and that it dated from the beginning of the Kwansei era (1789—1800). Hazama Jūfū was a pupil of Asada's, and was a merchant.

[1] It is said at about the age of forty.

[2] School of Instruction with Greatest Sincerity. It was also called the Sanwa-itchi school.

[3] His *Sanseki*, a collection of *tenzan* problems consists of 223 books.

[4] Endō, Book III, p. 57. On account of the importance of these schools in the history of education in Japan, the list is here reproduced for Western readers:

1. Momokawa Ryū, or Momokawa's School, teaching the *soroban* arithmetic as set forth in Momokawa's *Kameizan* of 1645.
2. Seki Ryū, or Seki's School.
3. Kūichi Ryū. The meaning is not known.
4. Nakanishi Ryū, or Nakanishi's School.
5. Miyagi Ryū, or Miyagi's School.
6. Takuma Ryū, or Takuma's School.
7. Saijō Ryū, or Superior School, sometimes incorrectly given as Mogami School.
8. Shisei Sanka Ryū, or Sanwa Itchi Ryū. The latter name may mean the Agreement of Trinity School.
9. Koryū, the Old School; or Yoshida Ryū, Yoshida's School.
10. Kurushima Gaku, or Kurushima's School.
11. Ōhashi Ryū, or Ōhashi's School.
12. Nakane Ryū, or Nakane's School, the Takebe-Nakane sect of the Seki School.
13. Nishikawa Ryū, or Nishikawa's School.
14. Asada Ryū, or Asada's School.
15. Hokken Ryū. The meaning is not known.

and when he died it was continued by his son, Furukawa Ken (1783—1837).

In this school, as in others of its kind, the *tenzan* algebra attracted much attention. It will be recalled that it was first made public in the *Shūki Sampō*, composed by Arima in 1769, a treatise written in Chinese characters and in such an obscure style as not easily to be understood. No better treatment appeared, however, until one was set forth by Sakabe Kōhan (1759—1824)[1] in 1810 under the title *Sampō Tenzan Shinan-Roku.*[2] In the same year two other works were written upon this subject, one by Ōhara Rimei[3] and the other by Aida,[4] but neither of these had the merit of Sakabe's treatise. Sakabe was in his younger days in the Fire Department of the shogunate, but he early resigned his post and became a *rōnin* or free *samurai*, devoting all of his time to study and to the teaching of his pupils. He first learned mathematics from Honda Rimei (1744—1821), who was a leader of the Takebe-Nakane sect of the Seki school, a man who was more of a patriot than a mathematician, but who knew something of the Dutch language and who was the first Japanese seriously to study the science of navigation from European sources. Sakabe also studied in the Araki-Matsunaga school and was one of the most distinguished pupils of Ajima. He left a noble record of a life devoted earnestly to the advance of his subject and to the assistance of his pupils.

---

16. Komura Ryū, or Komura's School, a school of surveying.
17. Furuichi Ryū, or Furuichi's School.
18. Mizoguchi Ryū, or Mizoguchi's School, a school of surveying.
19. Shimizu Ryū, or Shimizu's School, also a school of surveying.

[1] He was a prolific writer, his other more important works being the *Shinsen Tetsujutsu* (1795) and the *Kakujutsu-keimō* (Considerations on the theory of the polygon, 1802). These exist only in manuscripts. His literal name was Chūgaku.

[2] Exercise book on the tenzan methods.

[3] *Tenzan Shinan* (Exercises in the tenzan method). Ōhara died in 1831.

[4] *Sampō Tenshō-hō*, or *Sampō Tensei-hō*, Treatise on the *Tenshō* method. Aida called the *tenzan* method by the name *tenshō*.

Sakabe's treatise was published in fifteen Books, the last one appearing in 1815. One of the first peculiarities of the work that strikes the reader is the new arrangement of the *sangi*, which it will be recalled were differently placed for alternate digits by all early writers. Sakabe remarks that "it is ancient usage to arrange these sometimes horizontally and sometimes vertically, . . . but this is far from being a praiseworthy plan, it being a tedious matter to rearrange whenever the places of the digits are moved forwards or backwards." He adds: "I therefore prefer to teach my pupils in my own way, in spite of the ancient custom. Those who wish to know the shorter method should adopt this modern plan."

Sakabe classifies quadratic equations according to three types, much as such Eastern writers as Al-Khowarazmi and Omar Khayyam had done long before, and as was the custom until relatively modern times in Europe. His types were as follows:

$$-ax^2 + bx + c = 0,$$
$$ax^2 + bx - c = 0,$$
$$ax^2 - bx + c = 0,$$

and for these he gives rules that are equivalent to the formulas

$$x = \frac{1}{a}\left[\frac{b}{2} + \sqrt{\left(\frac{b}{2}\right)^2 + ac}\right],$$
$$x = \frac{1}{a}\left[-\frac{b}{2} + \sqrt{\left(\frac{b}{2}\right)^2 + ac}\right],$$

and

$$x_1 = \frac{1}{a}\left[\frac{b}{2} - \sqrt{\left(\frac{b}{2}\right)^2 - ac}\right],$$
$$x_2 = \frac{1}{a}\left[\frac{b}{2} + \sqrt{\left(\frac{b}{2}\right)^2 - ac}\right].$$

He takes, as will be seen, only the positive roots, neglecting the question of imaginaries, a type never considered in pure Japanese mathematics.[1]

[1] Seki knew that there are equations with no roots, the *mushō shiki* (equations without roots), but of the nature of the imaginary he seems to

Among his one hundred ninety-six problems is one in Book VI to find the smallest circle that can be touched internally by a given ellipse at the end of its minor axis, and the largest one that can be touched externally by a given ellipse at the end of its major axis. To solve the latter part he takes a sphere inscribed in a cylinder and cuts it by a plane through a point of contact, and concludes that the diameter of the maximum circle is $a^2 \div b$, where a is the minor axis and b is the major axis. For the other case he finds the diameter to be $b^2 \div a$.

Sakabe gives some attention to indeterminate equations. Thus in solving (Problem 104) the equation

$$2x^2 + y^2 = z^2$$

he takes any even number for x and separates $\frac{1}{2}x^2$ into two factors, m and n, then taking

$$y = m - n, \ z = m + n.$$

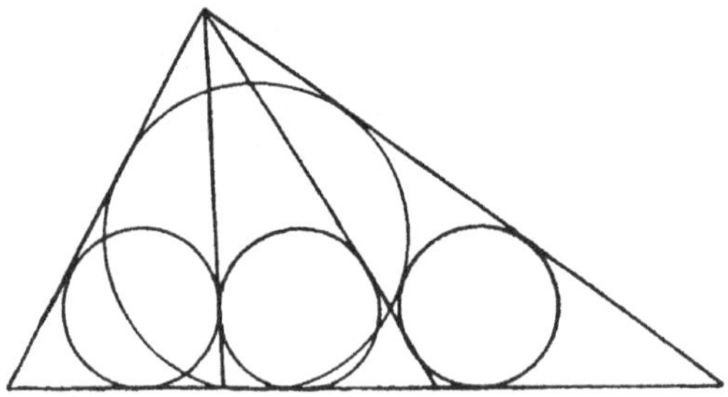

Among the geometric problems is the following (No. 138): "There is a triangle which is divided into smaller triangles by oblique lines so drawn from the vertex that the small inscribed circles as shown in the figure are all equal. Given the altitude h of the triangle and the diameter d of the circle inscribed

have been ignorant. In Kawai's *Kaishiki Shimpō* (1803) the statement is made that there may be a *mushō* (without root), that is, a root that is neither positive nor negative, but nothing is said as to the nature of such a root.

in the triangle, required to find the diameter of one of the n equal circles." His solution may be expressed by the formula

$$1-\frac{d}{h}=\left(1-\frac{x}{h}\right)^n,$$

where x is the required diameter.

In his Book X Sakabe gives some interesting methods of summing a series, but none that involved any principle not already known in Japan and in the world at large. They include the general plan of breaking simple series into partial geometric series, as in this case:

$$\begin{aligned}
s &= 1+2r+3r^2+4r^3+\cdots\cdots \\
&= 1+r+r^2+r^3+\cdots\cdots &&= \frac{1}{1-r} \\
&\quad\; +r+r^2+r^3+\cdots\cdots\cdots &&= \frac{r}{1-r} \\
&\quad\quad\; +r^2+r^3+\cdots\cdots\cdots &&= \frac{r^2}{1-r} \\
&\quad\quad\quad\; +r^3+\cdots\cdots\cdots\cdots &&= \frac{r^3}{1-r} \\
&= \frac{1}{1-r}+\frac{r}{1-r}+\frac{r^2}{1-r}+\frac{r^3}{1-r}+\cdots \\
&= \frac{1}{(1-r)^2}.
\end{aligned}$$

In the same way he sums

$$1+3r+6r^2+10r^3+15r^4+\cdots$$
$$1+4r+9r^2+16r^3+25r^4+\cdots$$
$$1+5r+14r^2+30r^3+55r^4+\cdots$$

and so on, these including the general types

$$\sum_0^\infty (i+1)^4 r^i,\qquad \sum_0^\infty (i+1)^5 r^i,\qquad \sum_0^\infty (i+1)^6 r^i,\cdots$$

$$\sum_{i=1}^{i=\infty}\left(\sum_{k=1}^{k=i} k^3\right) r^{i-1},\quad \sum_{i=1}^{i=\infty}\left(\sum_{k=1}^{k=i} k^4\right) r^{i-1},\cdots$$

In the extraction of roots Sakabe gives (Problem 167) a rule for the evaluation of $\sqrt[n]{N}$ that has some interest. He takes any number a_1 such that a_1^n is approximately equal to N. From this he obtains $a_2 = N \div a_1^{n-1}$. Then the real value of $\sqrt[n]{N}$ will evidently lie between a_1 and a_2, so that he takes for his third approximation $a_3 = \frac{1}{2}(a_1 + a_2)$, increasing or decreasing this slightly if it is known that $\sqrt[n]{N}$ lies nearer a_1 or a_2, respectively. He next calculates $a_4 = N \div a_3^{n-1}$, and continues this process as far as desired. Thus, to find $\sqrt[5]{0.125}$, let us take $a_1 = 0.66$. Then we find

$$a_2 = 0.6597541,$$
$$a_3 = 0.6597539553865$$

where a_2 is correct to 5 decimal places and a_3 to 12 decimal places.

Sakabe gives many other interesting problems, including various applications of the *yenri*. Among his results is the following series:

$$\frac{\pi}{4} = 1 - \frac{1}{5} - \frac{1\cdot 4}{5\cdot 7\cdot 9} - \frac{(1\cdot 3)\cdot(4\cdot 6)}{5\cdot 7\cdot 9\cdot 11\cdot 13} - \frac{(1\cdot 3\cdot 5)\cdot(4\cdot 6\cdot 8)}{5\cdot 7\cdots\cdots 15\cdot 17} - \cdots.$$

He also treats of the length of the arc in terms of the chord and altitude, as several writers had already done in the preceding century, and he was the first Japanese to publish rules for finding the circumference or an arc of the ellipse.[1]

Sakabe also wrote in 1803 a work entitled the *Rippō Eijiku*,[2] in which he treated of the cubic equation, the roots being expressed in a form resembling continued fractions which involved only square roots.[3] In 1812 he published his *Kwanki-*

---

[1] Ajima is doubtfully said to have discovered these rules, but he did not print them. Sakabe was the first to treat of the ellipse in a printed work.

[2] Or *Rippō Eichiku*. Literally, Methods of approximating by increase and decrease (the root of) a cubic.

[3] This work was never printed. The same plan had been attempted by one Fujita Seishin, of Tatebayashi in Jōshū, and his manuscript had been

kodo-shōhō,[1] a work on spherical trigonometry, and in 1816 his *Kairo-anshinroku*,[2] a work on scientific navigation.

The best-known of Sakabe's pupils was Kawai Kyūtoku,[3] a shogunate *samurai* of high rank and at one time a Superintendent of Finance. In 1803 Kawai published his *Kaishiki Shimpō*,[4] although it is suspected that Sakabe may have had a hand in writing it. He records in the preface that Sakabe had told how in his day some European and Chinese works had appeared in Japan, but that in none of them was found so general a method as he himself laid before his pupils. Indeed there was some truth in this boast, since the subject considered was the numerical higher equation, and, as we have seen, Horner's method had long been known in the East. It was here that China and Japan actually led the world, and when Sakabe and Kawai improved upon the work of their countrymen they *a fortiori* improved upon the rest of the mathematical fraternity.

This improvement consisted first in abandoning the *sangi* in favor of the *soroban*,[5] an ideal of all of the Japanese mathematicians of the eighteenth century. In the second place the general plan of work was simplified, as will be seen from the following summary of the process:

Let an equation of the nth degree, whose coefficients are integers, either positive or negative, be represented by

$$a_1 + a_2 x + a_3 x^2 + \cdots a_n x^{n-1} + a_{n+1} x^n = 0.$$

The n roots are generally positive or negative according as the pairs of coefficients (a_{n+1}, a_n), (a_n, a_{n-1}), $\cdots (a_2, a_1)$ have different signs or the same sign. The rth of these roots $(r = 1, 2, 3 \cdots n)$ may be found as follows:

submitted to Sakabe, who found it so complicated that he proceeded to simplify it in this work.

1 Literally, A short way to measure spherical arcs by the telescopic observation of heavenly bodies.

2 Literally, The safety of navigation.

3 Or Kawai Hisanori.

4 New method of solving equations.

5 See Kawai, *Kaishiki Shimpō* (1803); and Endō, Book III, p. 53.

First write

$$P = \cfrac{\cfrac{\cfrac{\frac{a_1}{A} + a_2}{A} + a_3}{\cdot \; \cdot \; \cdot \atop A} + a_{n-r+1}}{}$$

$$= \frac{a_1 + a_2 A + a_3 A^2 + \cdots + a_{n-r+1} A^{n-r}}{A^{n-r}}.$$

Then take

$$Q = a_{n-r+2} + a_{n-r+3} A + a_{n-r+4} A^2 + \cdots a_{n+1} A^{r-1},$$

and let $B = \frac{-P}{Q}$.

A may be assigned any value so long as P shall not have a different sign from a_{n-r+1} and Q shall not have a different sign from a_{n-r+2}.

Next proceed in the same way with A', denoting the result by B'.

If now we shall find either that

$$A > B \text{ and } A' < B'$$

or that

$$A < B \text{ and } A' > B',$$

then there will be in general a root of the equation between A and A'. Now by narrowing the limits between which the root lies a first approximation may be reached, but it suffices for a rough approximation to take the average of A, A', B and B'.

Repeat the same process with the first approximation as was followed with A and thus obtain a second approximation, and so on.

For example, take the equation

$$3360 - 2174x + 249x^2 - x^3 = 0.$$

Since a_3 and a_4 have different signs, the first root is positive. Let us begin with $A = 10$.

Then $\frac{3360}{10} = 336,$

$$336 - 2174 = -1838,$$

$$-\frac{1838}{10} = 183.8,$$

$$-183.8 + 249 = 65.2 = P.$$

Also $Q = -1,$

so that $B = -\frac{P}{Q} = 65.2.$

Similarly

$$A = 10 \qquad B = 65.2$$
$$A' = 100 \qquad B' = 227$$
$$A'' = 230 \qquad B'' = 239.6$$
$$A''' = 250 \qquad B''' = 240.3,$$

which shows that the first root lies between A'' and A''', since

$$A'' < B'' \text{ and } A''' > B'''.$$

Furthermore

$$\frac{A'' + A''' + B'' + B'''}{4} = 239.975, \text{ or nearly } 240,$$

which is the first approximation.

In the same way the approximate second root is 7.21. The rest of the computation is along lines previously known and already described.

In 1820 an architect named Hirauchi Teishin[1] published a work entitled *Sampō Hengyō Shinan*,[2] and later the *Shōka Kiku Yōkai*,[3] both intended for men of his profession and for engineers. Much use is made of graphic computation, as in the extraction of the cube root by the use of line intersections. In 1840 Hirauchi wrote another work, the *Sampō Chokujutsu Seikai*,[4] in which he treated of the geometric properties of

---

1 Also known by his earlier name of Fukuda Teishin.

2 Also transliterated *Sampō-Henkei-shinan*. Literally, Treatise on the Hengyō method, Hengyō meaning the changing of forms.

3 Literally, A short treatise on the line methods.

4 Exact notes on direct mathematical methods.

figures rather than of their mensuration. While the book had no special merit, it is worthy of note as being a step towards pure geometry, a subject that had been generally neglected in Japan, as indeed in the whole East.

It often happens in the history of mathematics, as in history in general, that some particular branch seems to show itself spontaneously and to become epidemic. It was so with algebra in medieval China, with trigonometry among the Arabs, with the study of equations in the sixteenth century Italian algebra, and with the calculus in the seventeenth century. So it was with the study of geometry in Japan. In the same year that Hirauchi brought out his first little work (1820), Yoshida Jūku published his *Kikujutsu Dzukai*[1] in which he attempted the solution of a considerable number of problems by the use of the ruler and compasses. It is true that this study had already been begun by Mizoguchi, and had been carried on by Murata Kōryū under whom Yoshida had studied, but the latter was the first of the Mizoguchi school[2] to bring the material together into satisfactory form.

About this time there lived in Ōsaka a teacher named Takeda Shingen, who published in 1824 his *Sampō Benran*,[3] in which the fan problems of the period appear (Fig. 42), and whose school exercised considerable influence in the western provinces. He also wrote the *Shingen Sampō*, a work that was published by his son in 1844. The old epigram which he adopted "There is no reason without number, nor is there number without reason," is well known in Japan.

It is, however, with the early stages of geometry that we are interested at this period, and the next noteworthy writer upon the subject was Hashimote Shōhō, who published his *Sampō Tenzan Shogakushō*[4] in 1830. The particular feature

---

[1] Illustrated treatise on the line method. His works are thought by some to have been written by Hasegawa.

[2] Endō, Book III, p. 91.

[3] Mathematical methods conveniently revealed. He is sometimes known by his familiar name, Tokunoshin.

[4] Tenzan method for beginners.

of interest in his work is the geometric treatment of the center of gravity of a figure. One of his problems is to find by geometric drawing the center of gravity of a quadrilateral, and the figure is given, although without explanation.[1]

This problem of the center of gravity now began to attract a good deal of attention in Japan. Perhaps the first real study[2] of the question was made by Takahashi Shiji, since a manuscript entitled *Tōkō Sensei Chojutsu Mokuroku*[3] mentions a work of his upon this subject. Since this writer was acquainted with the Dutch language and science, he doubtless received his inspiration from this source. His son Takahashi Keihō[4] (1786-1830) was, like himself, on the Astronomical Board of

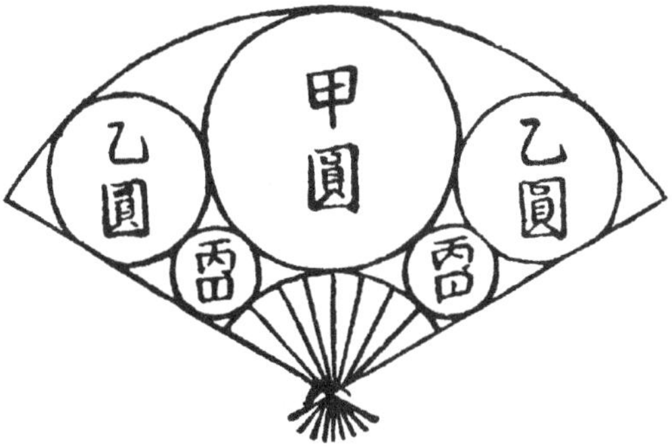

Fig. 42. From Takeda Shingen's *Sampō Benran* (1824).

the Shogunate, and was imprisoned from 1828 until his death in 1830, for exchanging maps with Siebold, whose work is mentioned in Chapter XIV.

Of the other minor writers of the opening of the nineteenth century the most prominent was Hasegawa Kan,[5] who published his *Sampō Shinsho* (New Treatise on Mathematics) in 1830

1 Endō, Book III, p. 107, gives a conjectural explanation. He is of the opinion that both the problem and the solution come from European sources.

2 The germ of the theory is found in Seki's writings.

3 List of Master Tōkō's writings, Tōkō being his *nom de plume*.

4 Or Takahashi Kageyasu.

5 Or Hasegawa Hiroshi.

under the name of one of his pupils. Hasegawa Kan was himself a pupil, and indeed the first and best-known pupil, of Kusaka Sei, the same who had studied under the celebrated Ajima, and hence he had good mathematical ancestry. His work was a compendium of mathematics, containing the *soroban* arithmetic, the "Celestial Element" algebra, the *tenzan* algebra, the *yenri*, and a little work on geometry, including some study of roulettes (Fig. 43). So well written was it that it became the most popular mathematical treatise in

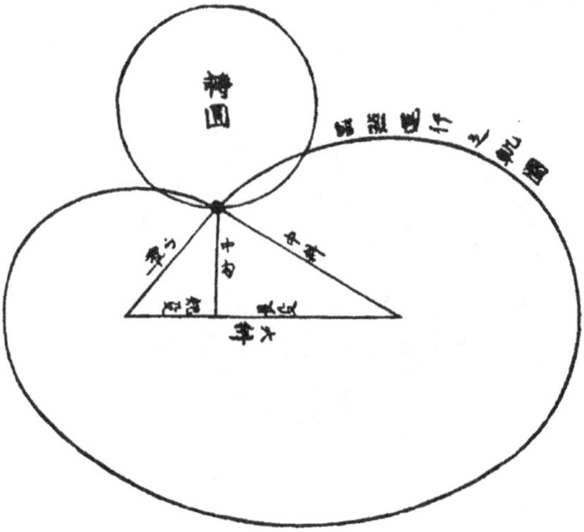

Fig. 43. From Hasegawa Kan's *Sampō Shinsho* (1849 edition).

the country and brought to its author much repute as a skilled compiler. Nevertheless the publication of this work led to great bitterness on the part of the Seki school, inasmuch as it made public the final secrets of the *yenri* that had been so jealously preserved by the members of this educational sect.[1] His act caused his banishment from among the disciples of Seki,[2] but it ended the ancient regime of secrecy

---

[1] The *yenri* here described is not the same as that of Ajima or Wada.

[2] Endō attributes his banishment to his having appropriated to his own use the money collected for printing Ajima's *Fukyū Sampō*.

in matters mathematical. Hasegawa died in 1838 at the age of fifty-six years.[1]

Among the noteworthy features of the *Sampō Shinsho* mention should be made of the reversion of series[2] in one of the geometric problems, and of the device of using limiting forms for the purpose of effecting some of the solutions. One of his algebraic-geometric problems is this: Given the diameters of the three escribed circles of a triangle to find the diameter of the inscribed circle. By considering the case in which the three escribed circles are equal, as one of the limits of form, Hasegawa gets on track of the general solution, a device that is commonly employed when we first consider a special case and attempt to pass from that to the general case in geometry. The principle met with severe criticism, it being obvious that we cannot reason from the square as a limit back to a rectangle on the one hand and a rhombus on the other. Nevertheless Hasegawa was very skilful in its use, and in 1835 he wrote another treatise upon the subject, the *Sampō Kyoku-gyō Shinan*,[3] published under the name of his pupil,[4] Akita Yoshiichi of Yedo.

It thus appears that the opening years of the nineteenth century were characterized by a greater infiltration of western learning, by some improvement in the *tenzan* algebra, and by the initial steps in pure geometry. None of the names thus far mentioned is especially noteworthy, and if these were all we should feel that Japanese mathematics had taken several steps backward. There was, however, one name of distinct importance in the early years of the century, and this we have reserved for a special chapter,—the name of Wada Nei.

---

1 Professor Hayashi gives the dates 1792-1832. But see Endō, Book II, p. 12, and Kawakita's article in the *Honchō Sūgaku Kōenshū*, p. 17.

2 An essentially similar problem, in connection with a literal equation of infinite degree, seems to have been first studied by Wada Nei.

3 Treatise on the method of limiting forms.

4 A custom of Hasegawa's. See the note on Hirauchi, above.

CHAPTER XII.

Wada Nei.

It will be recalled that in the second half of the eighteenth century Ajima added worthily to the *yenri* theory, bringing for the first time to the mathematical world of Japan a knowledge of a kind of integral calculus for the quadrature of areas and the cubature of volumes. The important work thus started by him was destined to be transmitted through his pupil, Kusaka Sei,[1] to a worthy successor of whom we shall now speak at some length.

Wada Yenzō Nei (1787-1840),[2] a *samurai* of Mikazuki in the province of Harima, was born in Yedo. His original name was Kōyama Naoaki, and in early life he served in Yedo in the Buddhist temple called by the name Zōjōji. He then changed his name for some reason, and is generally known in the scientific annals of his country as Wada Nei. After leaving the temple life he took up mathematics under the tutelage of Lord Tsuchimikado, hereditary calendar-maker to the Court of the Mikado. He first studied pure mathematics under a certain scholar of the Miyagi school, and then under Kusaka Sei. As has already been mentioned, this Kusaka compiled the *Fukyū Sampō* from the results of his contact with Ajima, thus bringing into clear light the teaching of his master. Although it must be confessed that he did not have the genius of Ajima, nevertheless Kusaka was a remarkable teacher,

---

[1] Endō, Book III, p. 127. See p. 172.

[2] Koide, *Yenri Sankyō*, preface. See Chapter XIV.

giving to mathematics a charm that fascinated his pupils and that inspired them to do very commendable work. Money had no attraction for him, and he lived a life of poverty, dying in 1839 at the age of seventy-five years.[1]

As to Wada, no book of his was ever published, and all of his large number of manuscripts, which were in the keeping of Lord Tsuchimikado, were consumed by fire,[2] that great and ever-present scourge of Japan that has destroyed so much of her science and her letters. Eking out a living by fortune-telling and by teaching penmanship, as well as by giving instruction in mathematics,[3] selling some of his manuscripts to gratify his thirst for liquor, Wada's life had little of happiness save what came as the reward of his teaching. He claimed to have had among his pupils some of the most distinguished mathematicians of his day,[4] men who came to him to learn in secret, recognizing his genius as an investigator and as a teacher.[5]

It will be recalled that Ajima had practiced his integration by cutting a surface into what were practically equal elements and summing these by a somewhat laborious process, and then passing to the limit for $n = \infty$. In a similar manner he found the volumes of solids. In every case some special series had to be summed, and it was here that the operation became tedious. Wada therefore set about to simplify matters by constructing a set of tables to accomplish the work of the modern table of integrals. Since his expression for "to integrate" was the Japanese word "to fold" (*tatamu*), these aids to calculation were called "folding tables" (*jō-hyō*), and of these he is known

---

1 ENDŌ, Book III, p. 121; C. KAWAKITA's article in the *Honchō Sūgaku Kōenshū*, p. 17; KOIDE, *Yenri Sankyō*, preface.

2 KOIDE, *Yenri Sankyō*, MS. of 1842, preface.

3 ENDŌ, Book III, p. 128.

4 The original list on some waste paper is now in the possession of N. Okamoto. The list includes the names of Shiraishi, Kawai, Uchida, Saitō, and Ushijima, with many others.

5 See also ENDŌ, Book III, p. 86.

to have left twenty-one, arranged in pamphlet form and bearing distinctive names.[1]

In 1818 Wada wrote the *Yenri Shinkō* in two books, published only in manuscript. In this he begins by computing the area of a circle in the following manner:

The diameter is first divided into $2n$ equal parts. Then, drawing the lines as shown in the figure, it is evident that

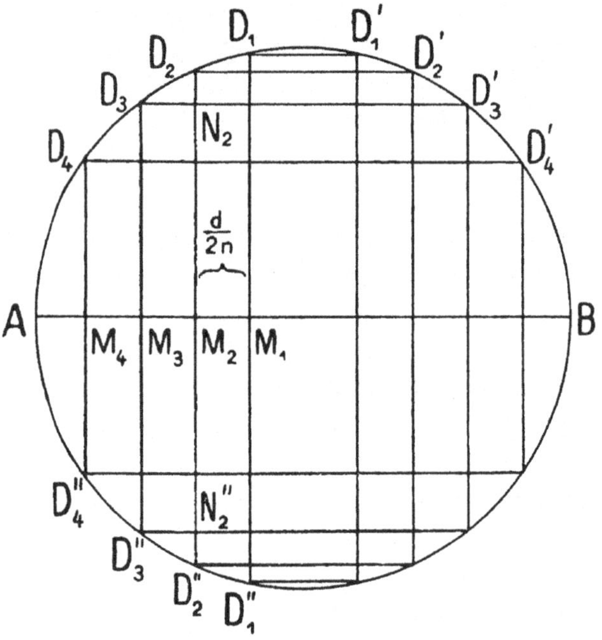

$$AM_{n-1} = M_{n-1}M_{n-2} = \cdots = \frac{d}{2n},$$

and

$$D_r D_r' = \frac{rd}{n},$$

whence

$$\overline{D_r D_r''}^2 = d^2 - \overline{D_r D_r'}^2$$

$$= d^2 - \frac{r^2 d^2}{n^2}.$$

---

[1] ENDO, Book III, p. 74.

Hence twice the area of $D_r D_r'' N_{r-1}'' N_{r-1}$

$$= D_r D_r'' \cdot \frac{d}{n} = \frac{d^2}{n} \sqrt{1 - \frac{r^2}{n^2}}$$

$$= \frac{d^2}{n} \left(1 - \frac{r^2}{2n^2} - \frac{1 \cdot r^4}{2 \cdot 4 n^4} - \frac{1 \cdot 3 r^6}{2 \cdot 4 \cdot 6 n^6} - \frac{1 \cdot 3 \cdot 5 r^8}{2 \cdot 4 \cdot 6 \cdot 8 r^8} - \cdots \right).$$

Summing for $r = 1, 2, 3, \ldots n$, we have

$$\frac{d^2}{n} \left(n - \frac{1}{2n^2} \sum_1^n r^2 - \frac{1}{2 \cdot 4 \cdot n^4} \sum_1^n r^4 - \cdots \right).$$

Multiplying, and then proceeding to the limit for $n = \infty$, we have the area of the circle expressed by the formula

$$a = d^2 \left(1 - \frac{1}{2 \cdot 3} - \frac{1}{2 \cdot 4 \cdot 5} - \frac{1 \cdot 3}{2 \cdot 4 \cdot 6 \cdot 7} - \frac{1 \cdot 3 \cdot 5}{2 \cdot 4 \cdot 6 \cdot 8 \cdot 9} - \cdots \right).$$

In the two operations of summing and proceeding to the limit Wada makes use of his "folding tables."

By a similar process Wada finds the circumference to be

$$2d \left(1 + \frac{1^2}{3!} + \frac{1^2 \cdot 3^2}{5!} + \frac{1^2 \cdot 3^2 \cdot 5^2}{7!} + \cdots \right),$$

and he obtains formulas for the area of a segment of a circle bounded by an arc and a chord, or by two arcs and two parallel chords.[1] It is also said that he gave upwards of a hundred infinite series expressing directly or indirectly the value of π,[2] among which were the following:

---

[1] For the complete treatment see HARZER, P., loc. cit., p. 33 of the Kiel reprint of 1905. HARZER shows that the formula used is essentially Newton's of 1666, given later by Wallis.

[2] ENDŌ, *A short account of the progress in finding the value of* π *in Japan* (in Japanese), in the *Rigakkai*, vol. III, No. 4, p. 24.

$$\frac{\pi}{2} = 1 + \frac{1}{3.2} + \frac{3}{5.8} + \frac{15}{7.48} + \frac{105}{9.384} + \frac{945}{11.3840} + \cdots$$

$$\frac{\pi}{4} = \frac{1}{3} + \frac{1}{5.2} + \frac{3}{7.8} + \frac{15}{9.48} + \frac{105}{11.384} + \frac{945}{13.3840} + \cdots$$

$$\frac{\pi}{8} = \frac{1}{3} + \frac{1}{15.2} + \frac{3}{35.8} + \frac{15}{63.48} + \frac{105}{99.384} + \frac{945}{143.3840} + \cdots$$

$$\frac{\pi}{32} = \frac{1}{15} + \frac{1}{35.2} + \frac{3}{63.8} + \frac{15}{99.48} + \frac{105}{143.384} + \cdots$$

$$\frac{\pi}{4} = 1 - \frac{1}{3} + \frac{3}{15} - \frac{15}{105} + \frac{105}{945} - \frac{945}{10395} + \cdots$$

$$\frac{\pi}{2\sqrt{2}} = 1 + \frac{1}{3.2.2} + \frac{3}{5.8.2^2} + \frac{15}{7.48.2^3} + \frac{105}{9.384.2^4} + \cdots,$$

the larger numbers in the denominators of these formulas being

2, 2.4, 2.4.6, ...

3, 3.5, 3.5.7, ...

1.3, 3.5, 5.7, ...

The same principle that he applies to the circle he also uses in connection with the ellipse,[1] finding the perimeter to be[2]

$$\pi a \left[1 - \sum_{1}^{\infty} \frac{1^2.3^2 \ldots (2n-3)^2 (2n-1)\, m^n}{1^2.2^2.\ldots \quad n^2}\right],$$

where $m = \frac{1}{4}\left(1 - \frac{b^2}{a^2}\right)$, and where for $n = 1$ the term is to be taken as $\frac{1}{1^2} m$.

Wada also turned his attention to the computation of volumes, simplifying Ajima's work on the two intersecting cylinders, and in general developing a very good working type of the integral calculus so far as it has to do with the question of mensuration.

The question of maxima and minima had already been considered by Seki more than a century before Wada's time, the

[1] In his *Setsu-kei Jun-gyaku*.

[2] Endō, Book III, p. 81.

rule employed being not unlike the present one of equating a differential coefficient to zero, although no explanation was given for the method. Naturally it had attracted the attention of many mathematicians of the Seki school, but no one had ventured upon any discussion of the reasons underlying the rule. The question is still an open one as to where Seki obtained the method. In the surreptitious intercourse with the West it would be just such a rule that would tend to find its way through the barred gateway, it being more difficult to communicate a whole treatise. At any rate the rule was known in the early days of the Seki school, and it remained unexplained for more than a century, and until Wada took up the question.[1] He not only gave the reason for the rule, but carried the discussion still further, including in his theory the subject of the maximum and minimum values of infinite series.[2] In this way he was able to apply the theory to questions involved in the *yenri* where, as we have seen, infinite series are always found.

In 1825 Wada wrote a work entitled *Iyen Sampō*[3] in which he treated of what he calls "circles of different species." He says that "if the area of a square be multiplied by the moment of circular area[4] it is altered[5] into a circle, and we have the area (of this circle). If the area of a rectangle be multiplied by the moment of circular area it is altered into an ellipse, and we have the area (of this ellipse). If the volume of a cube or a cuboid be multiplied by the moment of the spherical volume,[6] it is altered into a sphere or a spheroid, and we have its volume. These are processes that are well known. It is possible to generalize the idea, however, applying these

---

1 It is found in his manuscript entitled *Tekijin Hō-kyū-hō*.

2 Endō, Book III, p. 83.

3 On Circles of different species.

4 I. e., by $\frac{\pi}{4}$. We would say, $a = \pi r^2$. The Japanese, however, always considered the diameter instead of the radius.

5 This seems the best word by which to express the Japanese form.

6 I. e., by $\frac{4}{3}\pi$.

processes to the isosceles trapezium, to the rectangular pyramid, and so on, obtaining circles and spheres of different forms."

For example, given an ellipse inscribed in the rectangle $ABCD$ as here shown. Take YY' the midpoints of DC and AB, respectively and construct the isosceles triangle ABY.

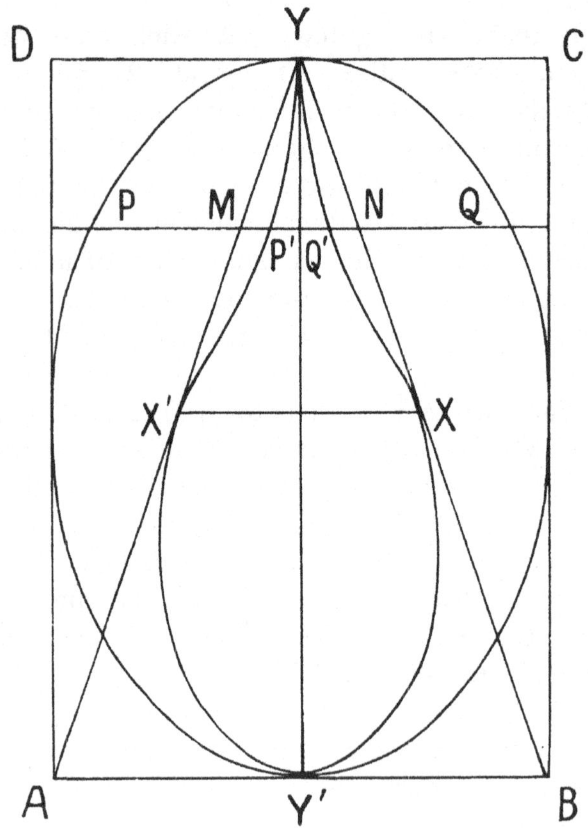

Draw any line parallel to AB cutting the ellipse in P and Q, and the triangle in M and N, as shown. Now take two points P', Q' on PQ, symmetric with respect to YY', and such that $AB : MN = PQ : P'Q'$. Then the locus of P' and Q' becomes a curve of the form shown in the figure, touching AY and BY at their mid-points X' and X, and the line AB

at Y'. If now we let $YY' = a$, and $X'X = b$, we may consider three species of curve,[1] namely for $a > b$, $a = b$, $a < b$.

Wada then finds the area inclosed by this curve to be $\frac{1}{4}\pi ab$, the process being similar to the one employed for the other curvilinear figures. He also generalizes the proposition by taking an isosceles trapezium instead of the isosceles triangle ABY, the area being found, as before, to be $\frac{1}{4}\pi ab$, where a and b are YY' and $X'X$ in the new figure.

Wada also devoted his attention to the study of roulettes, being the first mathematician in Japan who is known to have considered these curves. It is told how he one time hung before the temple of Atago, in Yedo, the results of his studies of this subject, although doing so in the name of one of his pupils. The problem and the solution are of sufficient interest to be quoted in substantially the original form.[2]

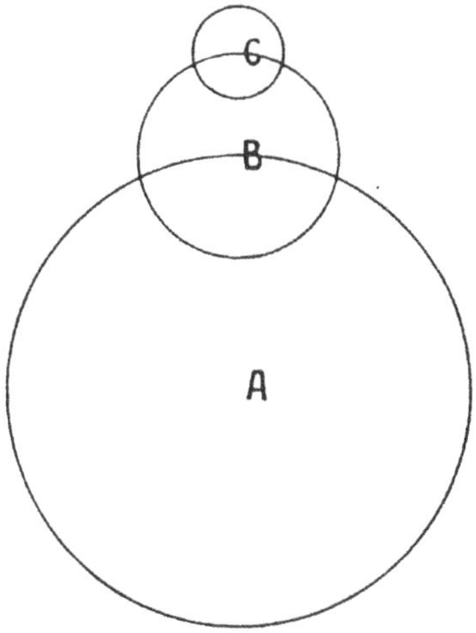

"There is a wheel with center A as in the figure, on the circumference of which is the center of a second wheel B, while on the circumference of B is the center of a third

[1] Wada calls these the *seitō-yen* (flourishing flame-shaped circle), *hōshu-yen*, and *suitō-yen* (fading flame-shaped circle).

[2] From the original. See also Endō, Book III, p. 103.

wheel, *C*. Beginning when the center *C* is farthest from the center *A*, the center *B* moves along the circumference of *A*, to the right, while the center *C* moves along the circumference of *B*, also to the right, the motions having the same angular velocity so that *C* and *B* return to their initial positions at the same time. Let the locus described by *C* be known as the *ki-yen* (the tortoise circle). Given the diameters of the wheels *A* and *B*, where the maximum of the latter should be half of the former, required to find the area of the *ki-yen*.

"Answer should be given according to the following rule: Take the diameter of the wheel *B*; square it and double; add the square of the diameter of *A*; multiply by the moment of the circular area, and the result is the area of the *ki-yen*.

"A pupil of Wada Yenzō Nei, the founder of new theories in the *yenri*, sixth in succession of instruction in the School of Seki."[1]

Wada's work in the domain of maxima and minima was carried on by a number of his contemporaries or immediate

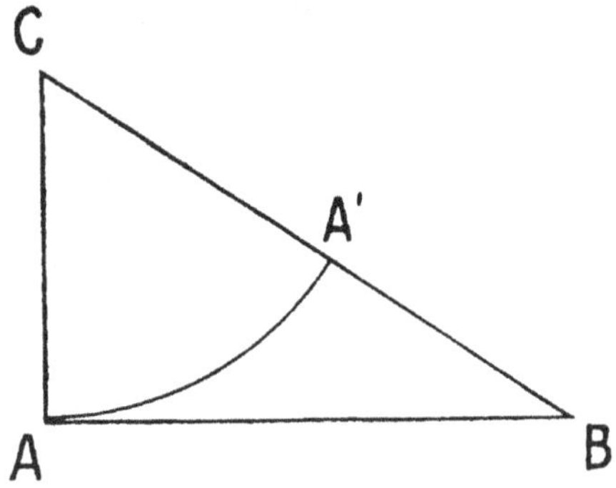

successors, among whom none did more for the theory than Kemmochi Yōshichi Shōkō. His contribution[2] to the subject is called the *Yenri Kyoku-sū Shōkai* (Detailed account of the

[1] The rule is equivalent to saying that the area is $\frac{1}{4}\pi(a^2 + 2b^2)$, where a and b are the diameters of *A* and *B*. Possibly this pupil was Koide Shūki. Wada's detailed solution is lost.

[2] Unpublished, and exact date unknown.

Circle-Principle method of finding Maxima and Minima), and contains two problems. The first of these problems is to find the shortest circular arc of which the altitude above its chord is unity. For this he gives two solutions, each too long to be given in this connection. His second problem is to construct a right triangle ABC with hypotenuse equal to unity, such that the arc AA' described with C' as a center, as in the figure, shall be the maximum, and to find the length of this maximum arc.[1]

---

[1] In KEMMOCHI's work there are certain transcendental equations which are solved by an approximation method known in Japan by the name *Kanrui-jutsu*, possibly due to Saitō Gigi or his father. Kemmochi certainly learned it from him. He also wrote a work usually attributed to Iwai Jūyen, the *Sampō yenri hio shaku*, one of the first to explain the *Kwatsu-jutsu* method.

It should be mentioned that the cycloid had been considered before Wada's time by Shizuki Tadao, who discussed it in his *Rekishō Shinsho* (1800).

CHAPTER XIII.

The Close of the Old Wasan.

Having now spoken of Wada's notable advance in the *yenri* or Circle Principle, in which he developed an integral calculus that served the ordinary purposes of mensuration, there remains a period of activity in this same field between the time in which he flourished and the opening of Japan to foreign commerce, which period marks the close of the old *wasan*, or native mathematics. Part of this period includes the labors of some of Wada's contemporaries, and part of it those of the next succeeding generation, but in no portion of it is there to be found a genius such as Wada. It was his work, his discoveries, his teaching that inspired two generations of mathematicians with the desire to further improve upon the Circle Principle. We have seen how the story is told that the best mathematicians of his day went to him in secret for the purpose of receiving instruction or suggestions, and it is further related that his range of discoveries was greater than his regular pupils knew, and that some of these discoveries appear as the work of others. This is mere rumor so far as any trustworthy evidence goes to show, but it lets us know the high estimate that was placed upon his abilities.

Among his contemporaries who gave serious attention to the *yenri* was a merchant of Yedo by the name of Iyezaki Zenshi who published a work in two parts, the *Gomei Sampō*, of which the first part appeared in 1814 and the second in 1826. There is a charming little touch of Japan in the fact that many of the problems relate to figures, and in particular to groups of ellipses, that can be drawn upon a folding fan, that is, upon a sector of an annulus.

Iyezaki gives also some problems in the *yenri* of a rather advanced nature. For example, he gives the area of the maximum circular segment that can be inscribed in an isosceles triangle of base b and so as to touch the equal sides s, as

$$\frac{(2s+b)s-b^2}{4\sqrt{s^2+\frac{b^2}{4}}}.$$

He also states that if an arc be described within a right triangle, upon the hypotenuse as the chord, and if a circle be drawn touching this arc and the two sides of the triangle, the maximum diameter of this circle is

$$\frac{1}{a}(a+b-c),$$

where a, b and c are the sides.

Contemporary with Iyezaki, or immediately following him, were several other writers who paid attention to figures drawn

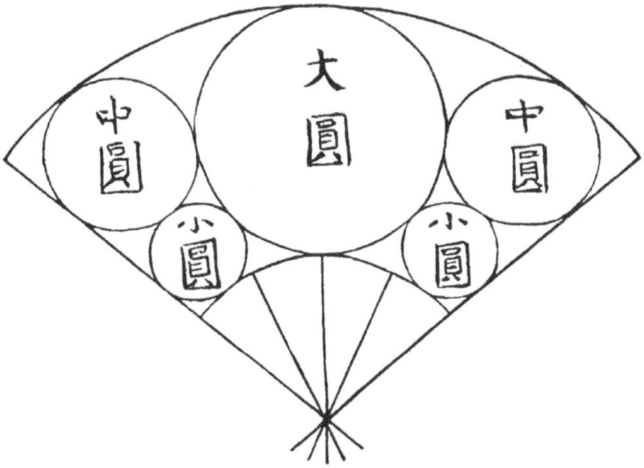

Fig. 44. From Yamada Jisuke's *Sampō Tenzan Shinan* (Bunkwa era, 1804—1818).

upon fans. Among these may be mentioned Yamada Jisuke whose *Sampō Tenzan Shinan* (Instructor in the *tenzan* mathematics) appeared early in the century (see Fig. 44); Takeda Tokunoshin whose *Kaitei Sampō* appeared in 1818 (see Fig. 45); Ishiguro Shin-yū (see Fig. 46), already mentioned in Chapter V as the last Japanese writer to make much of the practice of proposing problems for his rivals to solve; and Matsuoka

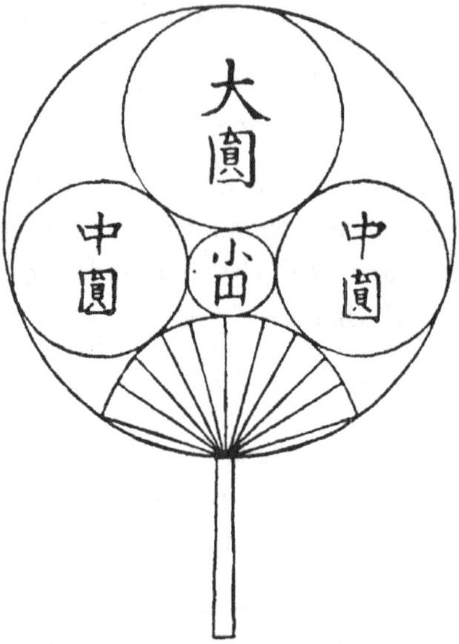

Fig. 45. From Takeda Tokunoshin's *Kaitei Sampō* (1818).

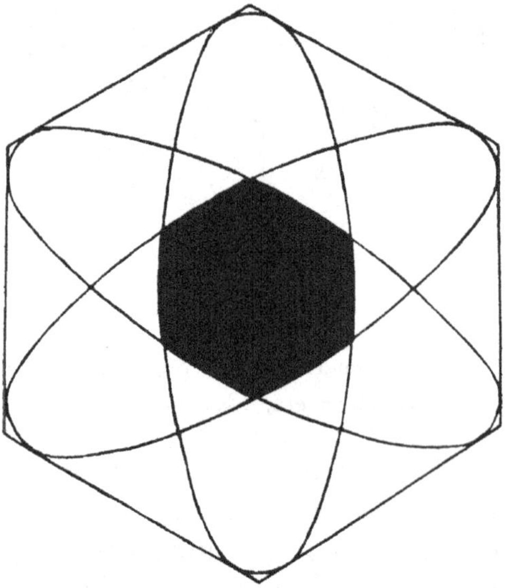

Fig. 46. Tangent problem from Ishiguro Shin-yū (1813).

Yoshikazu, whose *Sangaku Keiko Daizen*, an excellent compendium of mathematics, appeared in 1808 and again in 1849.

Also contemporary with Iyezaki was Shiraishi Chōchū (1796-1862) who published a work entitled *Shamei Sampu*[1] in 1826. He was a *samurai* in the service of Lord Shimizu, a near relative of the Shogun. While most of the problems in this treatise relate to the *yenri*, there is some interesting work in the line of indeterminate equations. One of these equations bears the name of Gokai Ampon, and like the rest was hung before some temple. The problem is as follows:

"There are three integral numbers, heaven, earth, and man, which being cubed and added together give a result of which the cube root has no decimal part. Required to find the numbers."

The problem is, of course, to solve the equation $x^3 + y^3 + z^3 = n^3$ in integers. The solution is given in Gokai's name, and he is known to have been an able mathematician, but whether it was his or Shiraishi's is unknown. In a manuscript commentary on the work[2] the following discussion of the equation appears:

First a table is constructed as follows:

$1^3 + 7 = 2^3$	$12^3 + 469 = 13^3$
$2^3 + 19 = 3^3$	$13^3 + 547 = 14^3$
$3^3 + 37 = 4^3$	$14^3 + 631 = 15^3$
$4^3 + 61 = 5^3$	$15^3 + 721 = 16^3$
$\boxed{5^3 + 91 = 6^3}$	$16^3 + 817 = 17^3$
$6^3 + 127 = 7^3$	$17^3 + 913 = 18^3$
$7^3 + 169 = 8^3$	$\boxed{18^3 + 1027 = 19^3}$
$8^3 + 217 = 9^3$	. . . . . .[3]
$9^3 + 271 = 10^3$	$\boxed{53^3 + 8587 = 54^3}$
$10^3 + 331 = 11^3$	. . . . .
$11^3 + 397 = 12^3$	$\boxed{102^3 + 31519 = 103^3}$

[1] Mathematical Results hung in Temples.

[2] *Shamei Sampu Kaigi.*

[3] In the table these missing numbers are given, but they are not necessary for our purposes.

Taking the second terms, 7, 19, 37, ..., it will be seen that the successive differences are as follows:

$$\begin{array}{ccccccccccc} 7 & & 19 & & 37 & & 61 & & 91 & & 127 \\ & 12 & & 18 & & 24 & & 30 & & 36 & \\ & & 6 & & 6 & & 6 & & 6 & & \end{array}$$

We can thus easily pick out the numbers that are the sums of two cubes, such as $91 = 3^3 + 4^3$, $1027 = 3^3 + 10^3$, and so on, and frame the corresponding relations as has been done in the table, adding others at will, such as

$$197^3 + 117019 = 198^3$$

$$306^3 + 281827 = 307^3.$$

Then writing $$n = y + 1,$$

from $$x^3 + y^3 + z^3 = n^3$$

we can derive

$$\frac{4(x^3 + z^3)}{3} - \frac{1}{3} = 4y^3 + 4y + 1. \qquad (1)$$

Then writing the selected equalities in the form

$$4^3 + 5^3 + 3^3 = 6^3 \qquad 31^3 + 102^3 + 12^3 = 103^3$$
$$10^3 + 18^3 + 3^3 = 19^3 \qquad 46^3 + 197^3 + 27^3 = 198^3$$
$$19^3 + 53^3 + 12^3 = 54^3 \qquad 64^3 + 306^3 + 27^3 = 307^3$$

we notice that our values of x, y, z, and n may be expressed as follows:

x	$3.1 + 1$	$3.3 + 1$	$3.6 + 1$	$3.10 + 1$	$3.15 + 1$	$3.21 + 1$
y	5	18	53	102	197	306
z	3.1^2	3.1^2	3.2^2	3.2^2	3.3^2	3.3^2
n	6	19	54	103	198	307

We therefore see that z is of the form $3a^2$. Corresponding to this value of z, x is of the form

$$x = 3(1 + 2 + 3 + \ldots + r) + 1,$$

where $r = 2a - 1$ or $2a$, alternately. That is,

$$x = 6a^2 \pm 3a + 1.$$

Substituting these values in (1) we have

$$324a^6 \pm 432a^5 + 360a^4 \pm 180a^3 + 60a^2 \pm 12a + 1$$
$$= 4y^2 + 4y + 1,$$

from which

$$y = 9a^3 + 6a^2 + 3a, \text{ or } 9a^3 - 6a^2 + 3a - 1,$$

and $n = y + 1 = 9a^3 + 6a^2 + 3a + 1$, or $9a^3 - 6a^2 + 3a$,

which gives the general solution.

Among the geometric problems given by Shiraishi two, given in Ikada's name, may be mentioned as types.

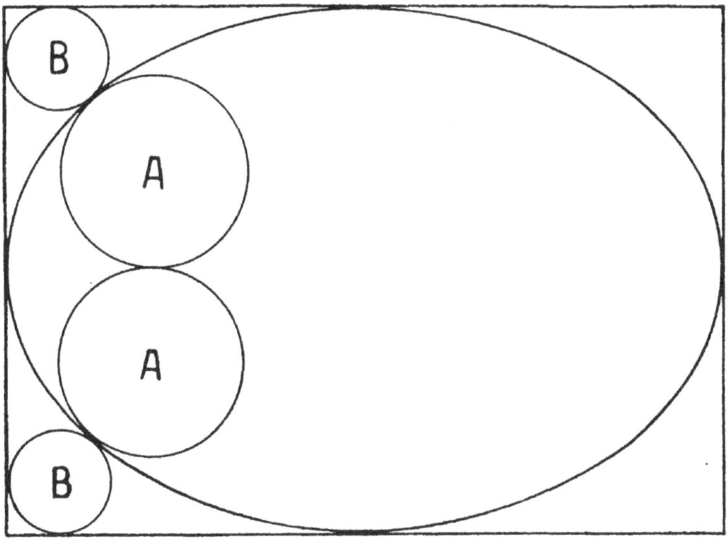

The first is as follows: "An ellipse is inscribed in a rectangle, and four circles which are equal in pairs are described as shown in the figure, A and B touching the ellipse at the same point. Given the diameters (a and b) of the circles, required to find the minor axis of the ellipse." The result is given as $a + b + \sqrt{(2a + b)\,b}$.

The second problem is to find the volume cut from a sphere by a regular polygonal prism whose axis passes through the center of the sphere.

There are also two problems given as solved by Shiraishi's pupils Yokoyama and Baishu, of which one is to find the volume cut from a cylinder by another cylinder that intersects it orthogonally and touches a point on the surface, and the other is to find the volume cut from a sphere by an elliptic cylinder whose axis passes through the center.

The *Shamei Sampu* contains a number of problems of this general nature, including the finding of the spherical surface that remains when a sphere is pierced by two equal circular cylinders that are tangent to each other in a line through the

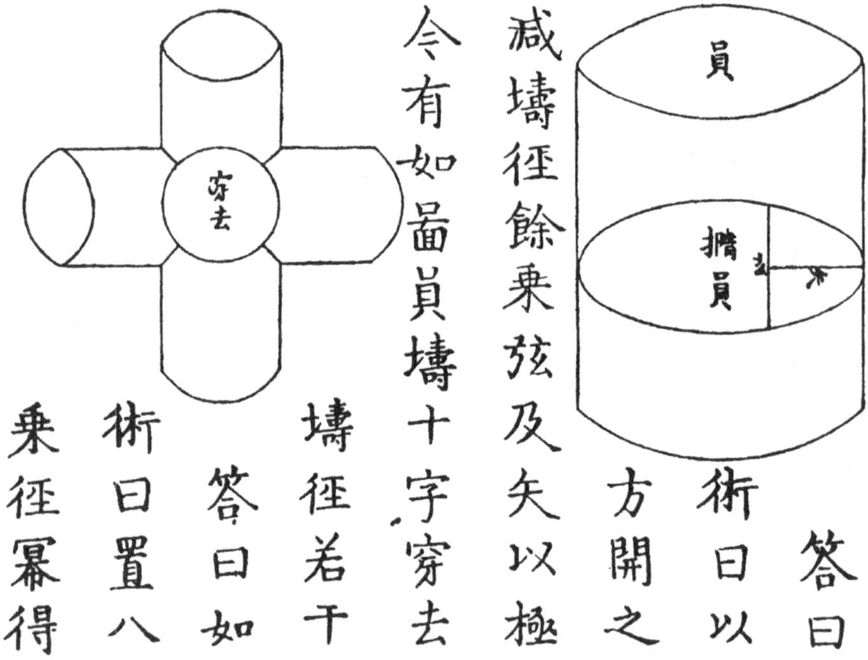

Fig. 47. From Iwai Jūyen's *Sampō Zasso* (1830).

center of the sphere; the finding of the area cut from a spherical surface by a cylinder whose surface is tangent to the spherical surface at one point; the finding of the volume cut from a cone pierced orthogonally to its axis by a cylinder, and the finding the surface of an ellipsoid.

Shiraishi also wrote a work entitled *Sūri Mujinzō*,[1] but it

[1] An inexhaustible Fountain of Mathematical Knowledge. It is given in Ikeda's name.

was never printed. It is a large collection of formulas and relations of a geometric nature. His pupil Kimura Shōju published in 1828 the *Onchi Sansō* which also contained

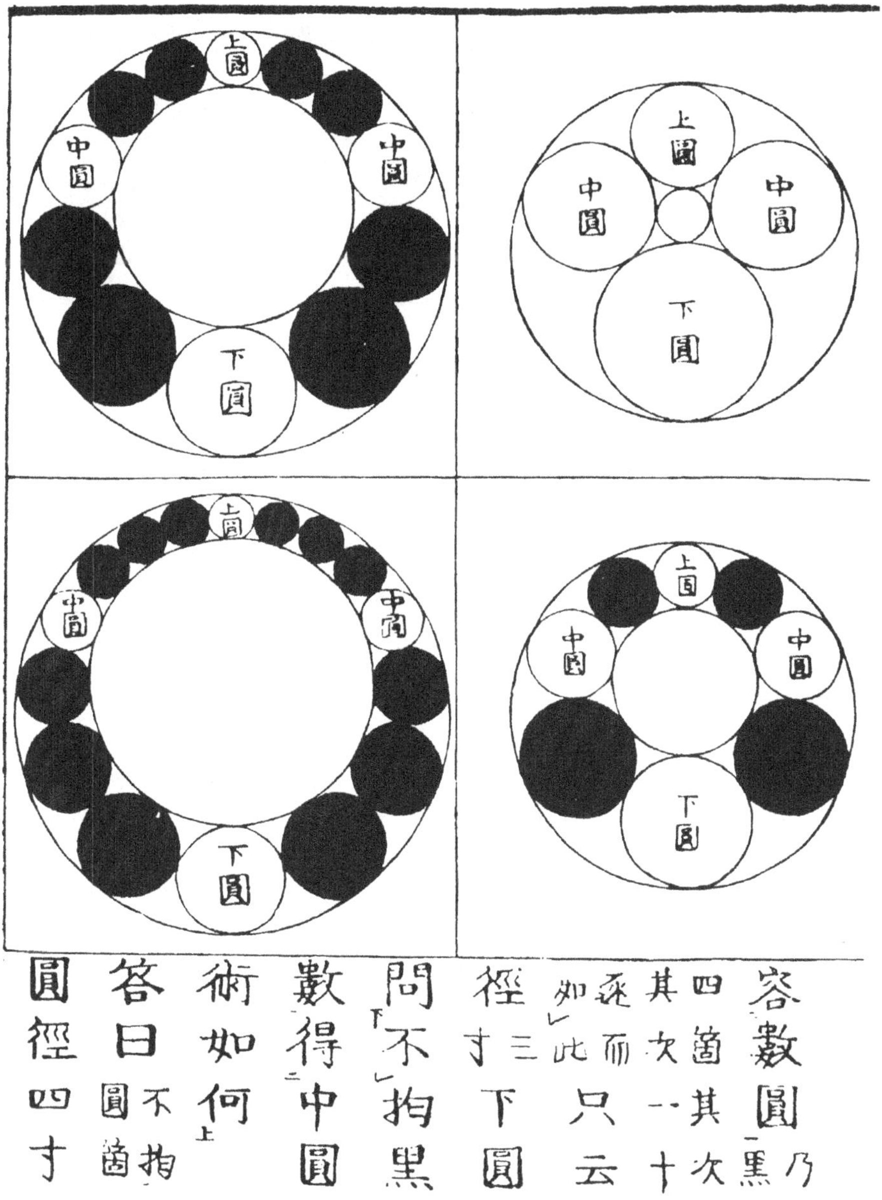

Fig. 48. From Aida Yasuaki's *Sampō Ko-kon Tsūran.*

numerous problems relating to areas and volumes. Interesting tangent problems analogous to those given by Shiraishi are found in numerous manuscripts of the nineteenth century. Illustrations are seen in Figs. 50 and 51, from an undated manuscript by one Iwasaki Toshihisa, and in Fig. 48, from a work by Aida Yasuaki.

Another work applying the *yenri* to mensuration, the *Sampō Zasso*, by Iwai Jūyen (or Shigetō), appeared in 1830. Iwai was a wealthy farmer living in the province of Jōshū and he had studied under Shiraishi. He also gives the problem of the intersecting cylinders (see Fig. 47), and the problem of finding the area of a plane section of an anchor ring. In

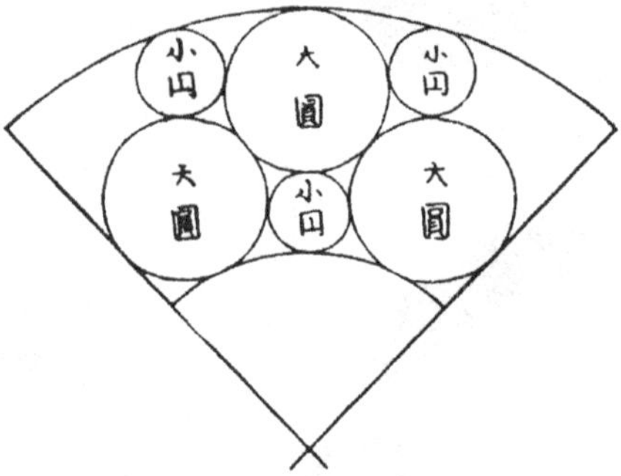

Fig. 49. From Hori-ike's *Yōmio Sampō* (1829).

1837 Iwai published a second work entitled *Yenri Hyōshaku*,[1] although it is said that this was written by Kemmochi Yōshichi. In this the higher order of operations of the *yenri* were first made public, and some notion of projection appears. Another work published in the same year, the *Keppi Sampō* by Hori-ike Hisamichi, resembles it in these respects. Hori-ike's *Yōmio Sampō* (1829) contains some interesting fan problems (see Fig. 49).

More talented as a mathematician, however, and much more popular, was Uchida Gokan,[2] who at the age of twenty-seven

---

1 The Method of the Circle Principle explained.

2 Or Uchida Itsumi.

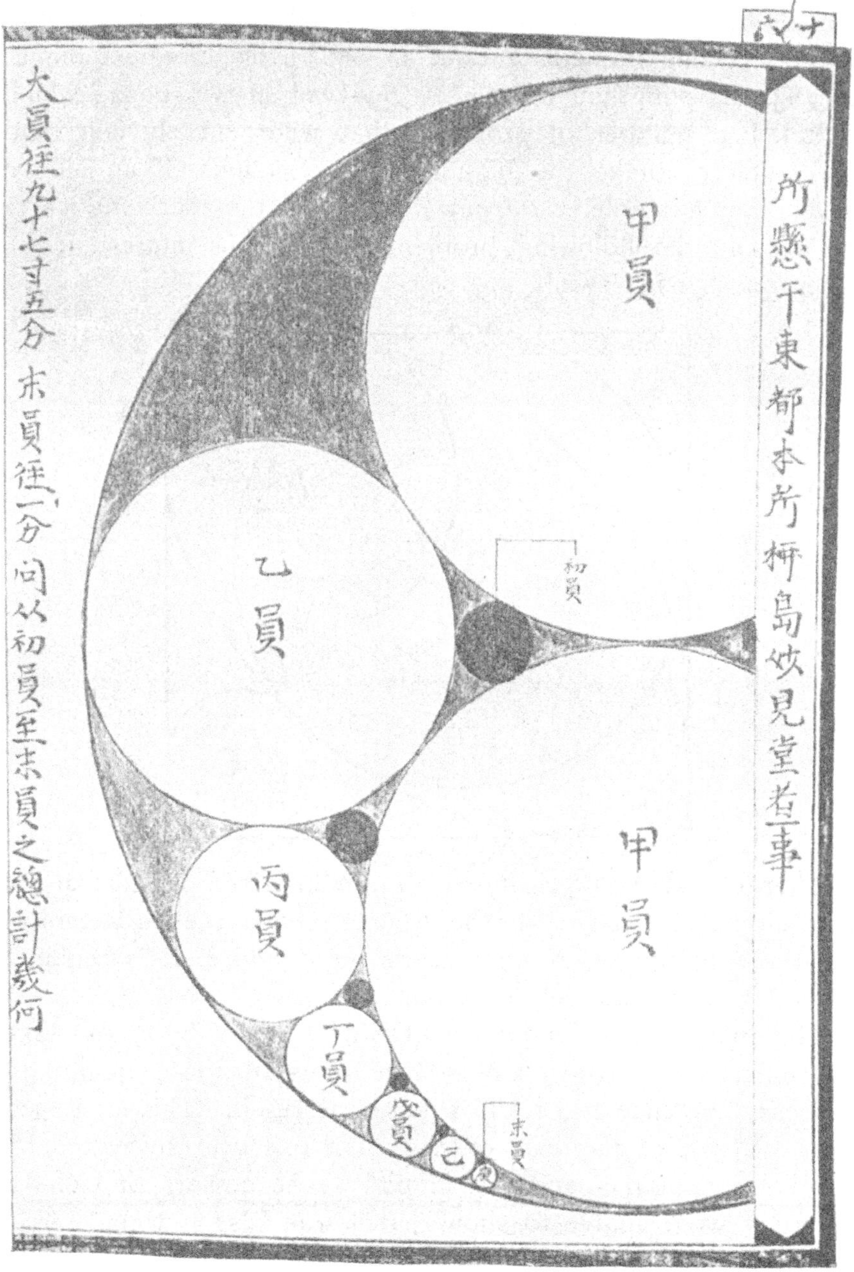

Fig. 50. Tangent problem, from a manuscript by Iwasaki Toshihisa.

published a work that brought him at once into prominence. Uchida was born in 1805 and studied mathematics under Kusaka, taking immediate rank as one of his foremost pupils. In 1832 he published his *Kokon Sankan*[1] in two books which included a number of problems that were entirely new, and did much to make the higher *yenri.* Sections of an elliptic wedge, for example, were new features in the mathematics of Japan, and the following problems showed his interest in the older questions as well:

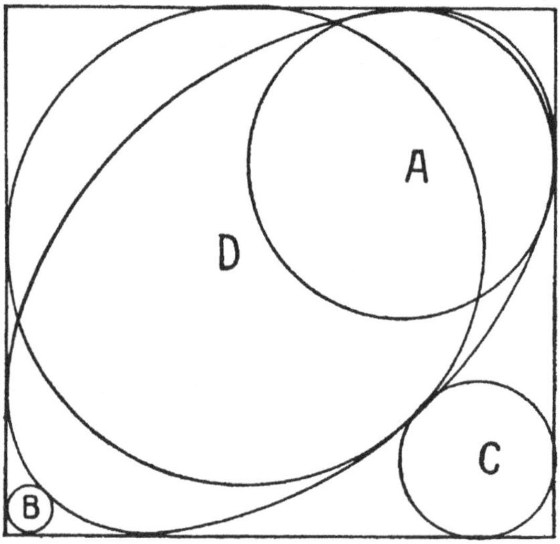

There is a rectangle in which are inscribed an ellipse and four circles as shown in the figure. Given the diameters of the three circles *A*, *B* and *C*, viz., *a*, *b* and *c*, it is required to find the diameter of the circle *D*.

The rule given is as follows: Divide *a* and *b* by *c*, and take the difference between the square roots of these quantities. To this difference add 1 and square the result. This multiplied by *c* gives the diameter of *D*. This rule was suspected by the contemporaries and the immediate successors of Uchida, but they were unable to show that it was false.[2] Uchida was,

1 Mirror (model) of ancient and modern Mathematical Problems.

2 For this information the authors are indebted to T. HAGIWARA, the only survivor, up to his death in 1909, of the leaders of the old Japanese school.

however, aware of it, although it appears in none of his writings.[1] Uchida also gave several interesting fan problems (see Fig. 55).

Uchida died in 1882, having contributed not unworthily to mathematics by his own writings, and also through the works of his pupils.[2] Among the latter works are Shino Chikyō's *Kakki Sampō* (1837), Kemmochi's *Tan-i Sampō* (1840)

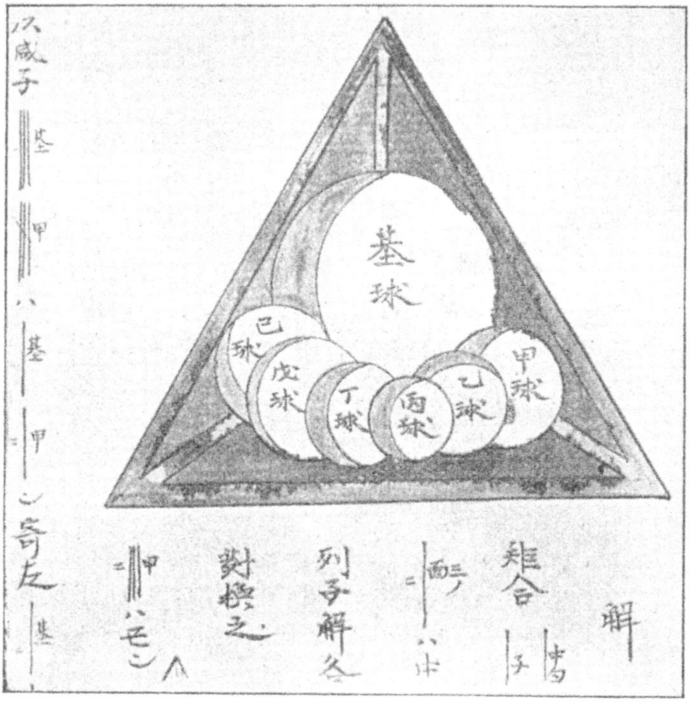

Fig. 51. Problem of spheres tangent to a tetrahedron, from a manuscript by Iwasaki Toshihisa.

and *Sampō Kaiwun* (1848), Fujioka's *Sampō Yenri-tsū* (1845), Takenouchi's *Sampō Yenri Kappatsu* (1849) and Kuwamoto Masaaki's *Sen-yen Kattsū* (1855), not to speak of several others.

[1] This information is communicated to us by C. KAWAKITA, one of Uchida's pupils.

[2] C. KAWAKITA's article in the *Honchō Sūgaku Kōen-shū*, 1908, p. 20. Shino Chikyō's *nom de plume* was Kenzan.

Among the contemporaries of Wada should also be mentioned Saitō Gigi, whose *Yenri-kan* appeared in 1834. It is possible that the real author was Saitō's father, Saitō Gichō (1784-1844), who also took much interest in mathematics. Father and son were both well-to-do farmers in Jōshū with whom mathematical work was more or less of a pastime. The *Yenri-kan* deserves this passing mention on account of the fact that it contains a problem on the center of gravity, and several problems on roulettes.

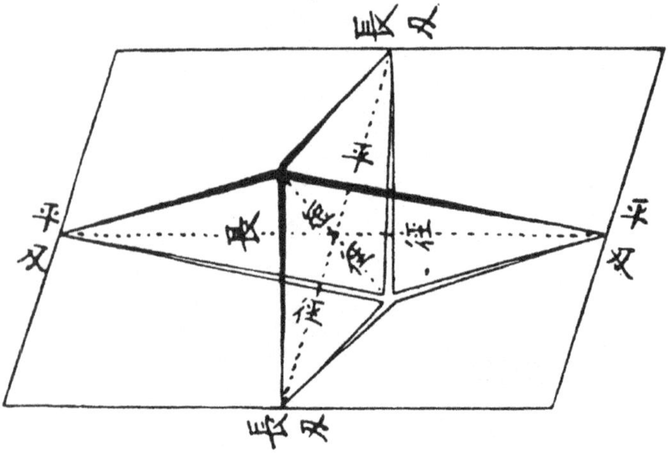

Fig. 52. From Kobayashi's *Sampō Koren* (1836).

In 1836 appeared Kobayashi Tadayoshi's *Sampō Koren* in which is considered the volumes of intersecting cylinders and a problem on a skew surface. The latter is stated as follows: "There is a 'rhombic rectangle'[1] which looks like a rectangle when seen from above, and like a rhombus when seen from the right or left, front or back. Given the three axes, required the area of the surface." Here the bases are gauche quadrilaterals. (The drawing is shown in Fig. 52.) Saitō also published a similar work, the *Yenri Shinshin,* in 1840.

[1] This is the literal translation of *choku bishi.* The figure is a solid and is defined in the problem.

At about the same period there appeared numerous works of somewhat the same nature, of which the following may be mentioned as among the best:

Gokai Ampon's (1796–1862) *Sampō Semmon Shō* (1840), a work on the advanced *tenzan* theory, with some treatment of magic squares (Fig. 54).

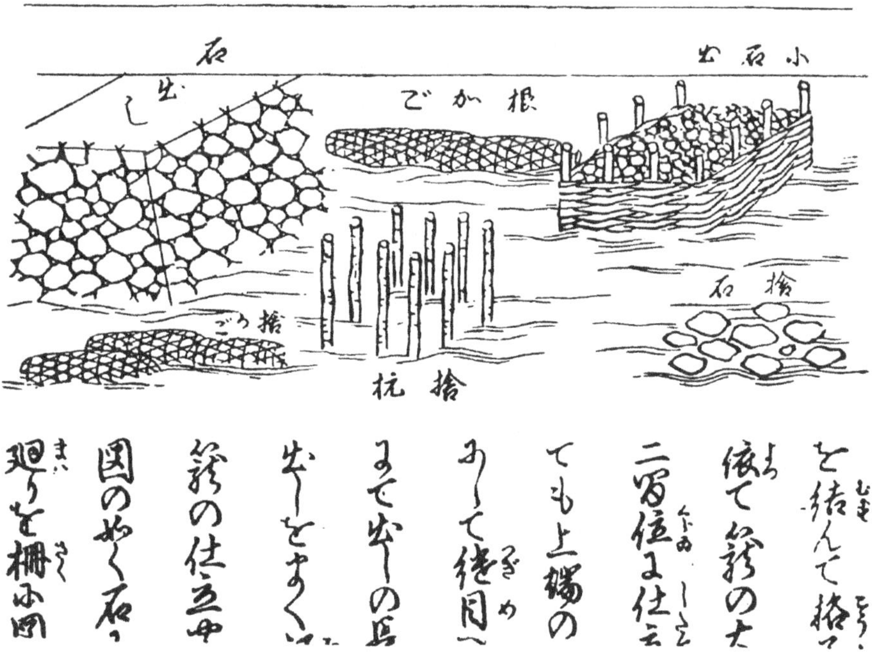

Fig. 53. From Murata's *Sampō Jikata Shinan* (1835).

Yamamoto Kazen's *Sampō Jojutsu*[1] (1841), containing an extensive list of formulas and excellent illustrations of the problems of the day (see Fig. 57).

Murata Tsunemitsu's *Sokuyen Shōkai* (1833), relating to the *tenzan* algebra applied to the ellipse, and his *Sampō Jikata Shinan* (1835), dealing with enginering problems (Fig. 53). Murata's pupil Toyota wrote the *Sampō Dayen-kai* in 1842, also relating to the *tenzan* algebra applied to the ellipse.[2]

[1] Aids in Mathematical Calculation.

[2] Besides Murata's work we have consulted Endō, Book III, p. 129.

Fig. 54. Magic Squares from Gokai's *Sampō Semmon Shō* (1840).

A work by a Buddhist priest, Kakudō written in Kyōto in 1794 and published in 1836, entitled *Yenri Kiku Sampō*, giving a summary of the *yenri*.

Chiba Tanehide's *Sampō Shin-sho* (1830), a large compendium of mathematics, actually the work of Hasegawa Kan.

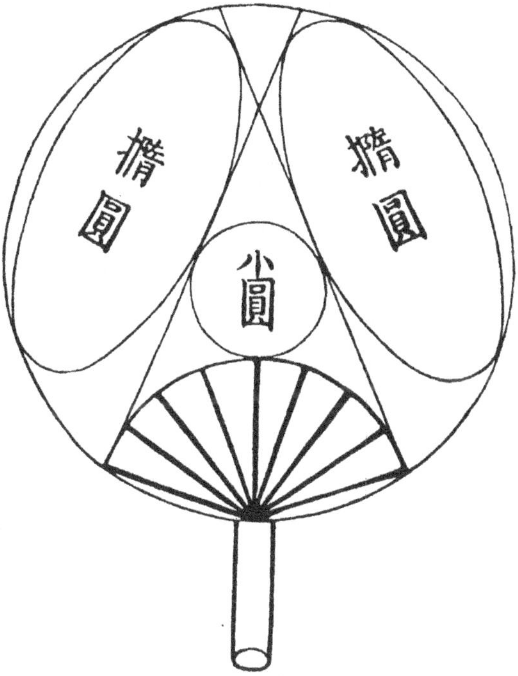

Fig. 55. From Uchida's *Kokon Sankan* (1832).

The *Sampō Tenzan Tebikigusa*, of which the first part was published by Yamamoto in 1833 and the second part by Ōmura Isshū (1824—1891) in 1841. This was a treatise on

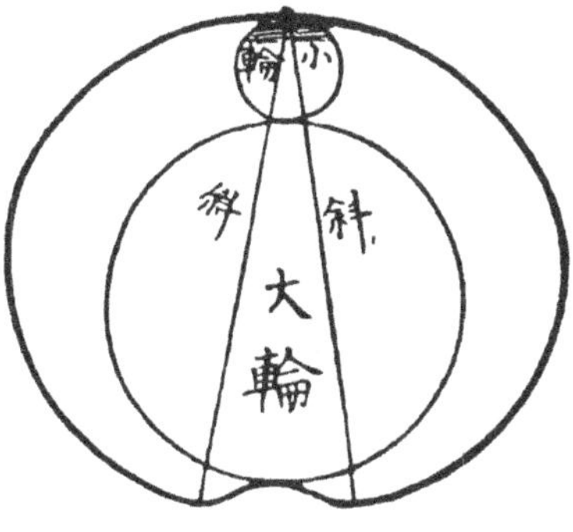

Fig. 56. From Minami's *Sampō Yenri Sandai* (1846).

tenzan algebra. Some of the fan problems in this work are of considerable interest. (See Fig. 58.)

Kikuchi Chōryō's *Sampō Seisū Kigenshō* (1845), a treatise on indeterminate analysis.

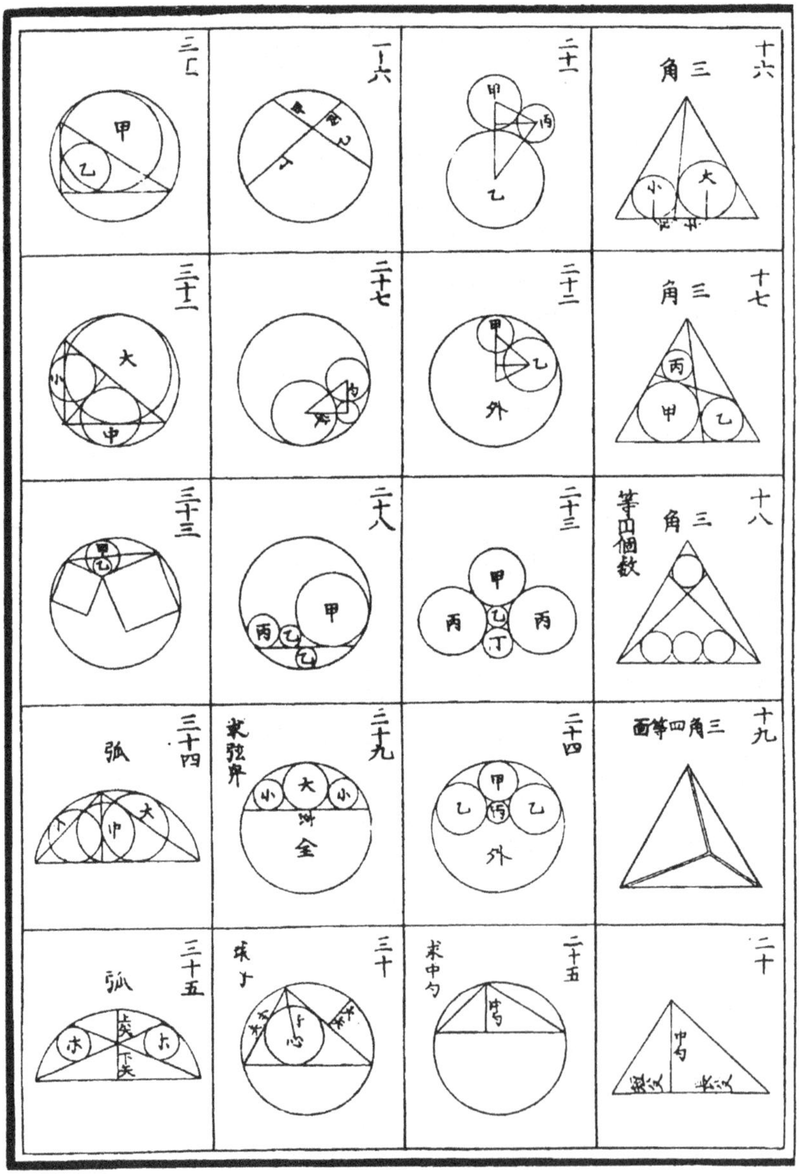

Fig. 57. From Yamamoto Kazen's *Sampō Jojutsu* (1841).

Minami Ryōhō's *Sampō Yenri Sandai* (1846), with some treatment of roulettes (see Fig. 56) and the *Juntendō Sampu*[1] (1847) by Iwata Seiyo and Kobayashi (not Tadayoshi). Curiously, the first ten pages of Minami's work are numbered with Arabic numerals.

Kaetsu's *Sampō Yenri Katsunō* (1851), a work on the higher *yenri*. This was considered of such merit that it was reprinted in China.

Iwasaki Toshihisa's *Yachu zak kai* (1831), *Saku yen riu kwai*

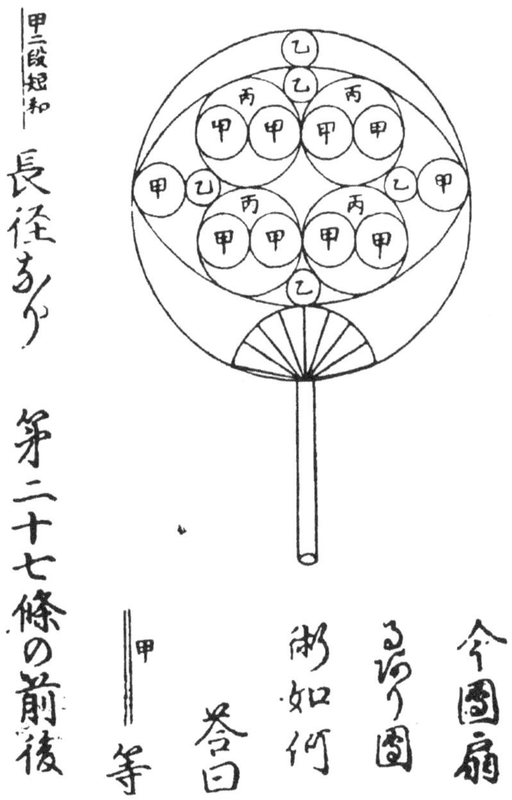

Fig. 58. From Yamamoto and Ōmura Isshū's Sampō Tenzan Tebikigusa (1833, 1841).

[1] *Juntendō* Mathematical Problems.

gi, and *Shimpeki sampō*, all works of considerable merit in the line of geometric problems.

Baba Seito's *Shi-satsu Henkai* (1830), generally known by the later title *Sampō Kishō*.

Hasegawa Kō's *Kyūseki Tsūkō*[1] (1844), published under the name of his pupil Uchida Kyūmei. This is more important than the works just mentioned. It consists of five books and gives a very systematic treatment of the *yenri*, beginning with the theory of limits and the use of the "folding tables" of Wada Nei. It treats of the circular wedge and its sections, of the intersections of cylinders and spheres (see Fig. 59), of ovals, or circles of various classes, as studied by Wada, and also of the cycloid and epicycloid.

The study of the catenary begins about 1860. The first to give it attention were Ōmura and Kagami, but the first printed work in which it is discussed is the *Sampō Hōyen-kan* (1862) of Hagiwara Teisuke (1828—1909). Another interesting problem which appears in this work is that of the locus of the point of contact of a sphere and plane, the sphere rolling around on the plane and always touching an anchor ring that is normal to and tangent to the plane. Hagiwara also published a work entitled *Sampō Yenri Shiron* (1866) in which he corrected the results of thirty-four problems given in twenty-two works published at various dates from the appearance of Arima's *Shūki Sampō* (1769) to his own time (see Figs. 60, 61). He also published a work entitled *Yenri San-yō* (1878), the result of his studies of the higher *yenri* problems. His manuscript called the *Reikan Sampō* was published in 1910 through the efforts of a number of Japanese scholars. Hagiwara was born in 1828, and was a farmer in narrow circumstances in the province of Jōshū. Not until about 1854 did he take an interest in mathematics, but when he recognized his taste for the subject he became a pupil of Saitō's, traveling on foot ten miles on the eve of a holiday so as to have a full day with his teacher. His manuscripts were horded in a miserly fashion

---

[1] General Treatment of Quadrature and Cubature.

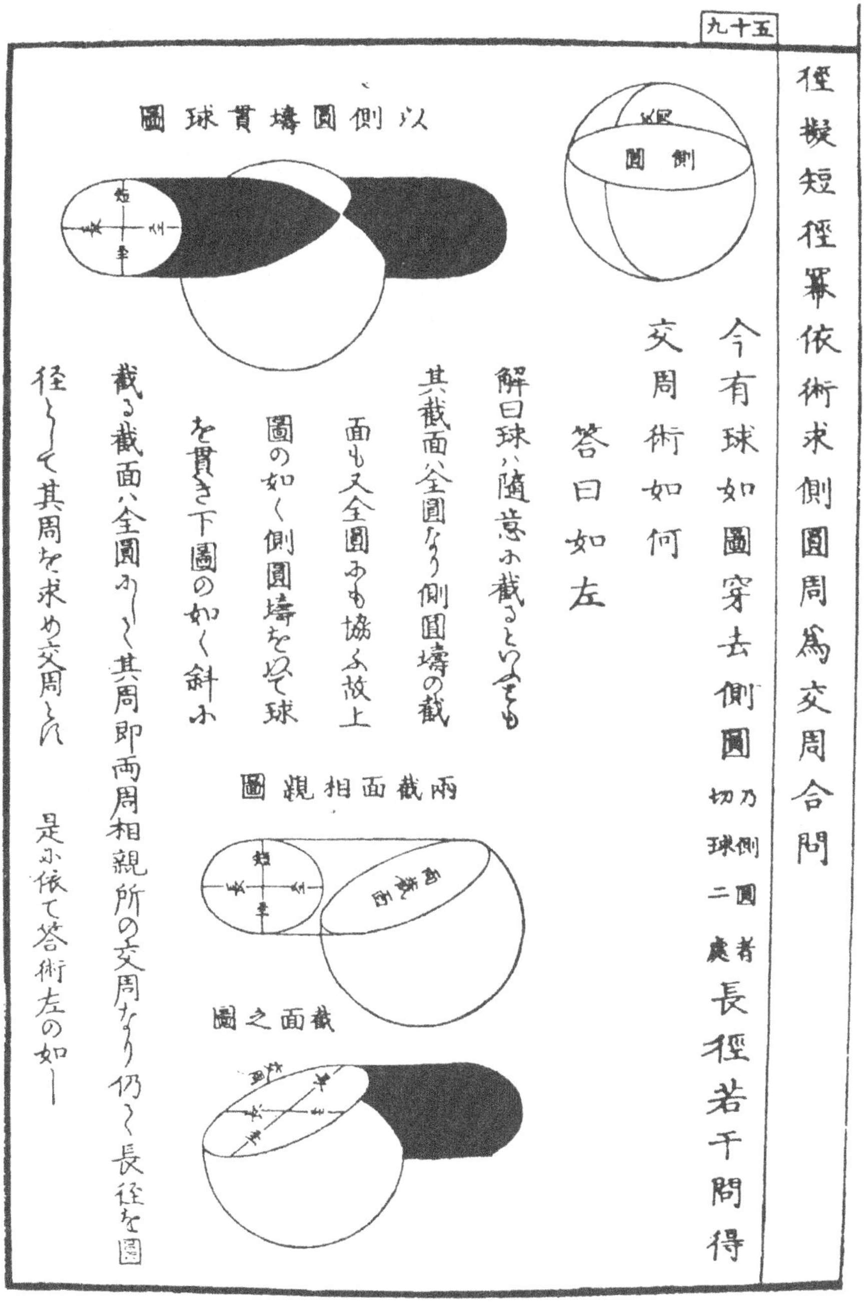

Fig. 59. From Hasegawa Kō's Kyūseki Tsūkō (1844).

until his death, November 28, 1909, when the last great mathematician of the old school passed away.

Mention should be made at this time of the leading mathematicians who were the contemporaries of Hagiwara, and who were living when the Shogunate gave place to the Empire in 1868. Of these, Hōdōji Wajūrō was born in 1820 and died in 1871.[1] He was the son of a smith in Hiroshima, and although he led a kind of vagabond existence he had a good deal of mathematical ability. It is said that he was the real author of Kaetsu's *Yenri Katsunō*. Several other books are known to have been written by him, but they were not published under his own name.

Iwata Kōsan (1812—1878), born a *samurai*, devoted his attention particularly to the ellipse. The following is his best known problem:

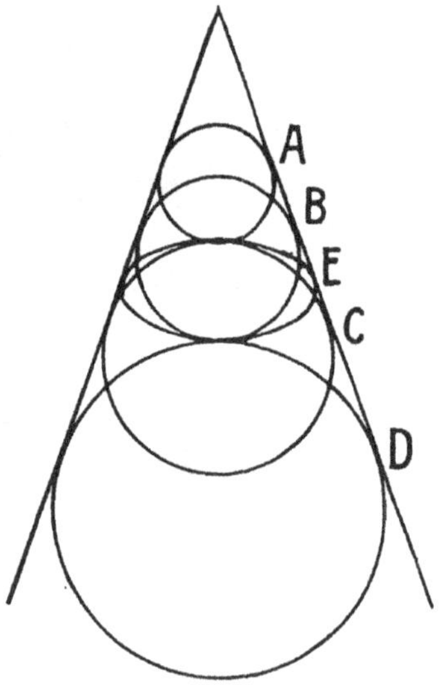

Given an ellipse E tangent to two straight lines and to four circles, A, B, C, D, as shown in the figure. Given the diameters of A, B and C, required to find the diameter of D. His solution, given in 1866, is essentially the proportion $a:b = c:d$, where a, b, c, d are the respective diameters of A, B, C and D. The problem was afterwards extended to any four conics instead of four circles, by H. Terao and others.

Kuwamoto Masaaki wrote the *Senyen Kattsū* in 1855, and in it he treated of roulettes of various kinds (see Fig. 62), of elliptic wedges (see Fig. 63), and other forms at that time attracting attention.

Takaku Kenjirō (1821—1883) wrote the *Kyokusū Taisei-jutsu* in which he made some contribution to the theory of maxima and minima.

1 C. Kawakita, in the *Honchō Sūgaku Kōenshū*, p. 23.

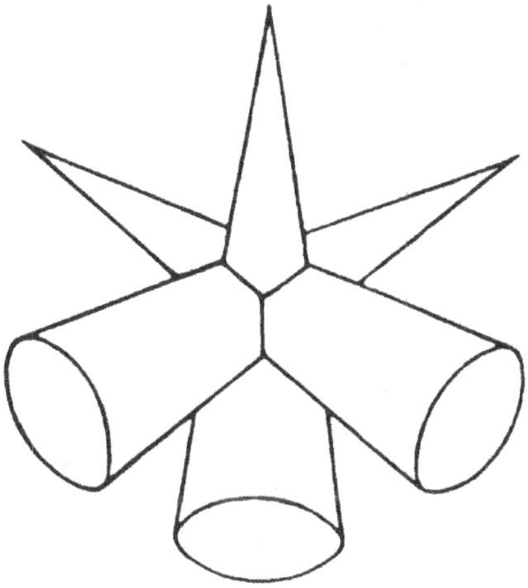

Fig. 60. From Hagiwara's *Sampō Yenri Shiron* (1866).

Fukuda Riken (1815—1889) lived first in Ōsaka and finally in Tokyō. He was a teacher of some prominence, and his *Sampō Tamatebako* appeared in 1879.

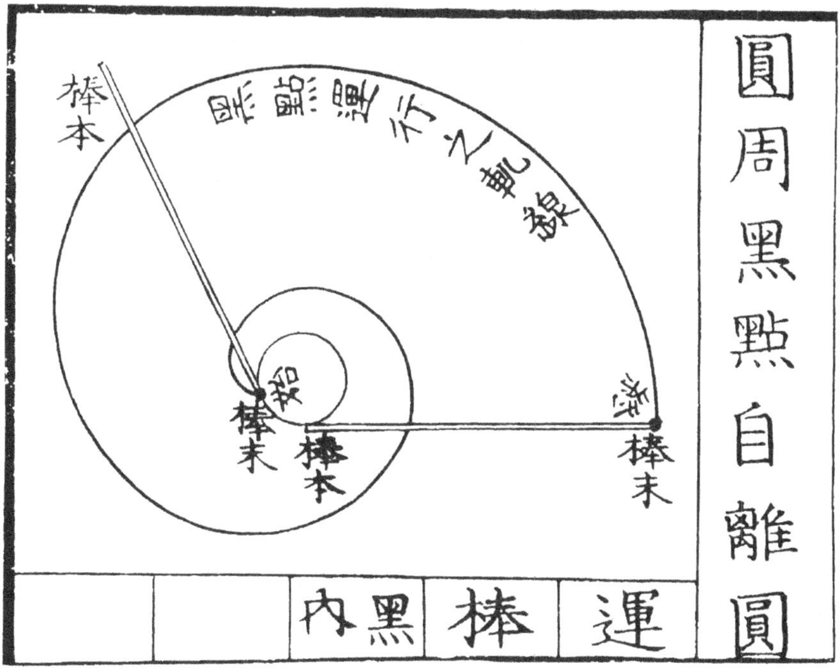

Fig. 61. From Hagiwara's *Sampō Yenri Shiron* (1866).

Yanagi Yūyetsu (1832—1891) was a naval officer who gave some attention to the native Japanese mathematics.

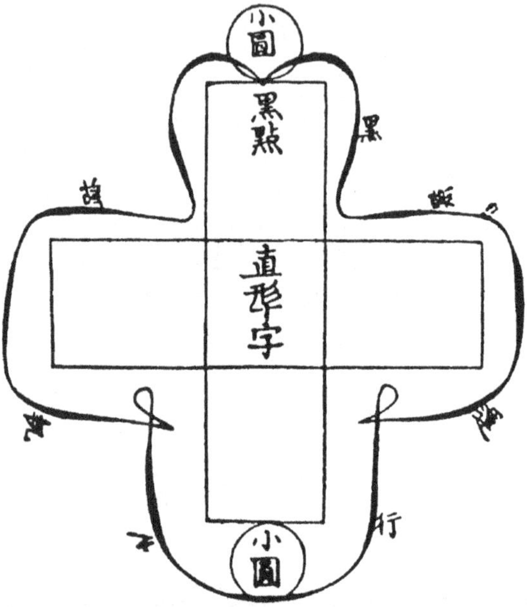

Fig. 62. From Kuwamoto Masaaki's *Sen yen Kattsū* (1855).

Suzuki Yen, who may still be living wrote a work (1878) upon circles inscribed in or circumscribed about figures of various shapes.

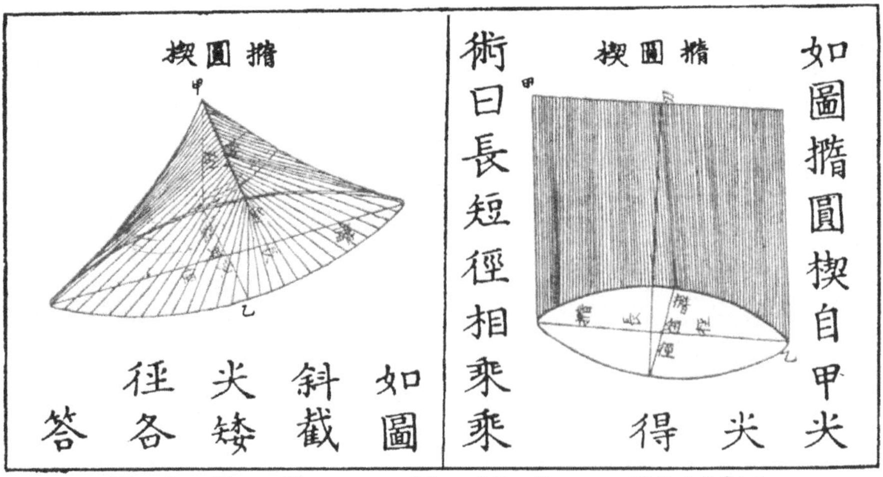

Fig. 63. From Kuwamoto Masaaki's *Sen yen Kattsū* (1855).

Thus closes the old *wasan,* the native mathematics of Japan. It seems as if a subconscious feeling of the hopelessness of the contest with Western science must have influenced the last half century preceding the opening of Japan. There was really no worthy successor of Wada Nei in all this period, and the feeling that was permeating the political life of Japan, that the day of isolation was passing, seems also to have permeated scientific circles. With the scholars of the country obsessed with this feeling of hopelessness as to the native mathematics, the time was ripe for the influx of Western science, and to this influence from abroad we shall now devote our closing chapter.

CHAPTER XIV.

The Introduction of Occidental Mathematics.

We have already spoken at some length in Chapter IX of the possible connection, slight at the most, between the mathematics of Japan and Europe in the seventeenth century. The possibility of such a connection increased as time went on, and in the nineteenth century the mathematics of the West finally usurped the place of the *wasan*. During this period of about two centuries, from 1650 to the opening of Japan to the world, knowledge of the European mathematics was slowly finding its way across the barriers, not alone through the agency of the Dutch traders at Nagasaki, but also by means of the later Chinese works which were written under the influence of the Jesuit missionaries. These missionaries were men of great learning, and they began their career by impressing this learning upon the Chinese people of high rank. Matteo Ricci (1552—1610), for example, with the help of one Hsü Küang-ching (1562—1634), translated Euclid into the Chinese language in 1607, and he and his colleagues made known the Western astronomy to the savants of Peking. It must be admitted, however, that only small bits of this learning could have found a way into Japan. Euclid, for example, seems to have been unknown there until about the beginning of the eighteenth century, and not to have been well known for two and a half centuries after it appeared in Peking.

Some mention should, however, be made of the work done for a brief period by the Jesuits in Japan itself, a possible influence on mathematics that has not received its due share of attention.[1] It is well known that the wreck of a Portuguese vessel upon the shores of Japan in 1542 led soon after to the efforts of traders and Jesuit missionaries to effect an entry into the country. In 1549 Xavier, Torres, and Fernandez landed at Kagoshima in Satsuma. Since in 1582 the Japanese Christians sent an embassy carrying gifts to Rome, and since it was claimed about that time that twelve thousand[2] converts to Christianity had been received into the Church, the influence of these missionaries, and particularly that of the "Apostle of the Indies," St. Francis Xavier, must have been great. In 1587 the missionaries were ordered to be banished from Japan, and during the next forty years a process of extermination of Christianity was pursued throughout the country.

In none of this work, not even in the schools that the Jesuits are known to have established in Japan, have we a definite trace of any instruction in mathematics. Nevertheless the influence of the most learned order of priests that Europe then produced, a priesthood that included in its membership men of marked ability in astronomy and pure mathematics, must have been felt. If it merely suggested the nature of the mathematical researches of the West this would have been sufficient to account for some of the renewed activity of the seventeenth century in the scientific circles of Japan. That the influence of the missionaries on mathematics was manifested in any other way than this there is not the slightest evidence.

It should also be mentioned that an Englishman named William Adams lived in Yedo for some time early in the seventeenth century and was at the court of Iyeyasu. Since he gave instruction in the art of shipbuilding and received honors at court, his opportunity for influencing some of the practical mathematics of the country must be acknowledged. There is also extant in a manuscript, the *Kikujutsu Denrai no Maki*, a story that one Higuchi Gonyemon of Nagasaki, a

---

[1] There is only the merest mention of it in P. HARZER's *Die exakten Wissenschaften im alten Japan*, Kiel, 1905.

[2] Some even claimed 200,000, at least a little later. E. BOHUM, *Geographical Dictionary*, London, 1688.

scholar of merit in the field of astronomy and astrology, learned the art of surveying from a Dutchman named Caspar, and not only transmitted this knowledge to his people but also constructed instruments after the style of those used in Europe. Of his life we know nothing further, but a note is added to the effect that he died during the reign of the third Shogun (1623—1650). A further note in the same manuscript relates that from 1792 to 1796 a certain Dutchman, one Peter Walius (?) gave instruction in the art of surveying, but of him we know nothing further.

In the eighteenth century the possibility that showed itself in the seventeenth century became an actuality. European sciences now began to penetrate into Japanese schools, either directly or through China. In the year 1713, for example, the elaborate Chinese treatises, the *Li-hsiang K'ao-ch'êng* and the *Su-li Ching-Yün,* which had been compiled by Imperial edict, were published in Peking. Of these the former was an astronomy and the latter a work on pure mathematics, and each showed a good deal of Jesuit influence. These books were early taken to Japan, and thus some of the trend of European science came to be known to the scholars of that country. There was also sent across the China Sea the *Li-suan Ch'üan-shu* in which Mei Wen-ting's works were collected, so that Japanese mathematicians not only came into some contact with Europe, but also came to see the progress of their science among their powerful neighbors of Asia. Takebe, for example, is said to have studied Mei's works and to have written some monographs upon them in 1726.[1]

Nakane Genkei (1662—1733) also wrote, about the same time, a trigonometry and an astronomy (see Fig. 64) based on the European treatment,[2] the result certainly of a study of Mei Wen-ting's works and possibly of the *Su-li Ching-Yün.*

---

[1] ENDŌ, Book II, p. 69. There is a copy in the Imperial Library.

[2] The *Hassen-hyō Kaigi* (Notes on the Eight Trigonometric Lines), and the *Tenmon Zukwai Hakki* (1696). He also wrote the *Kōwa Tsūreki* and the *Ko reki Sampō* (1714).

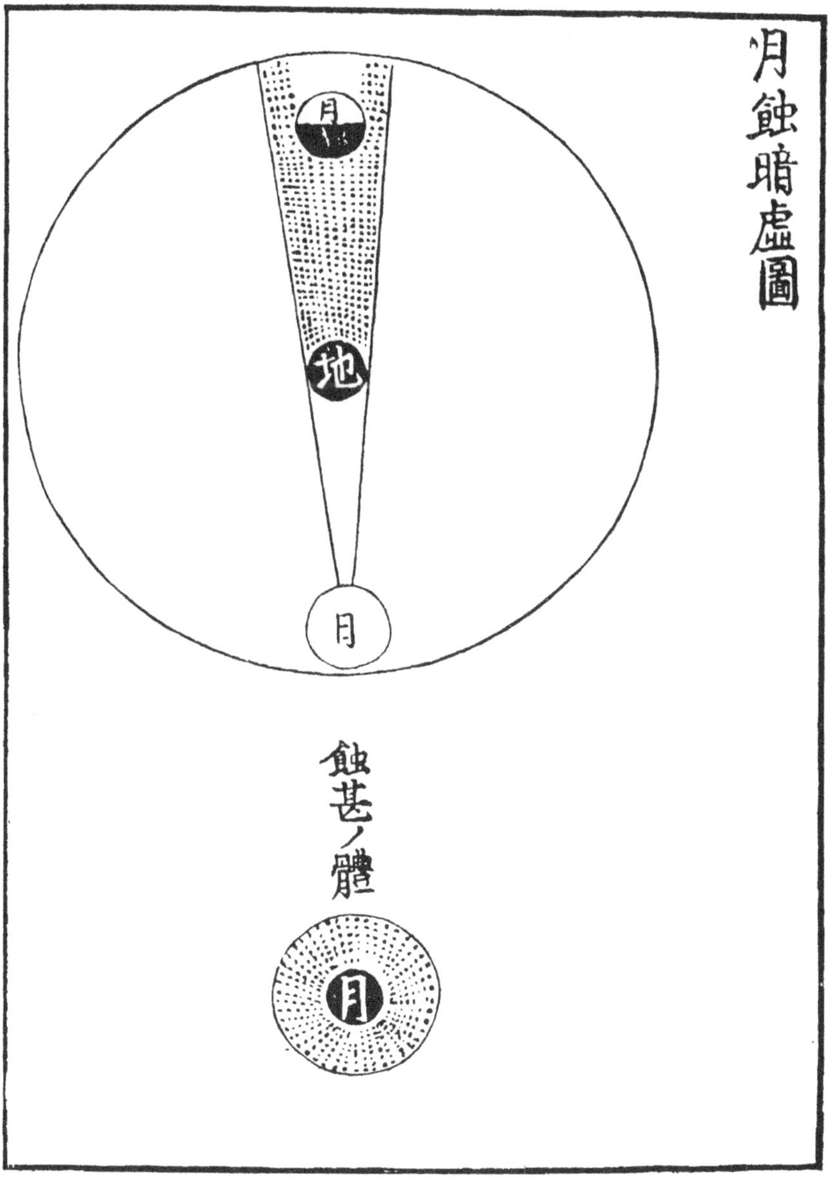

Fig. 64. From Nakane Genkei's astronomy of 1696.

His pupil Kōda Shin-yei, who died in 1758, also wrote upon the same subject. The illustrations given from the works on surveying by Ogino Nobutomo in his *Kiku Genpō Chōken* of 1718 (Fig. 65), and Murai Masahiro in his *Riochi Shinan* of

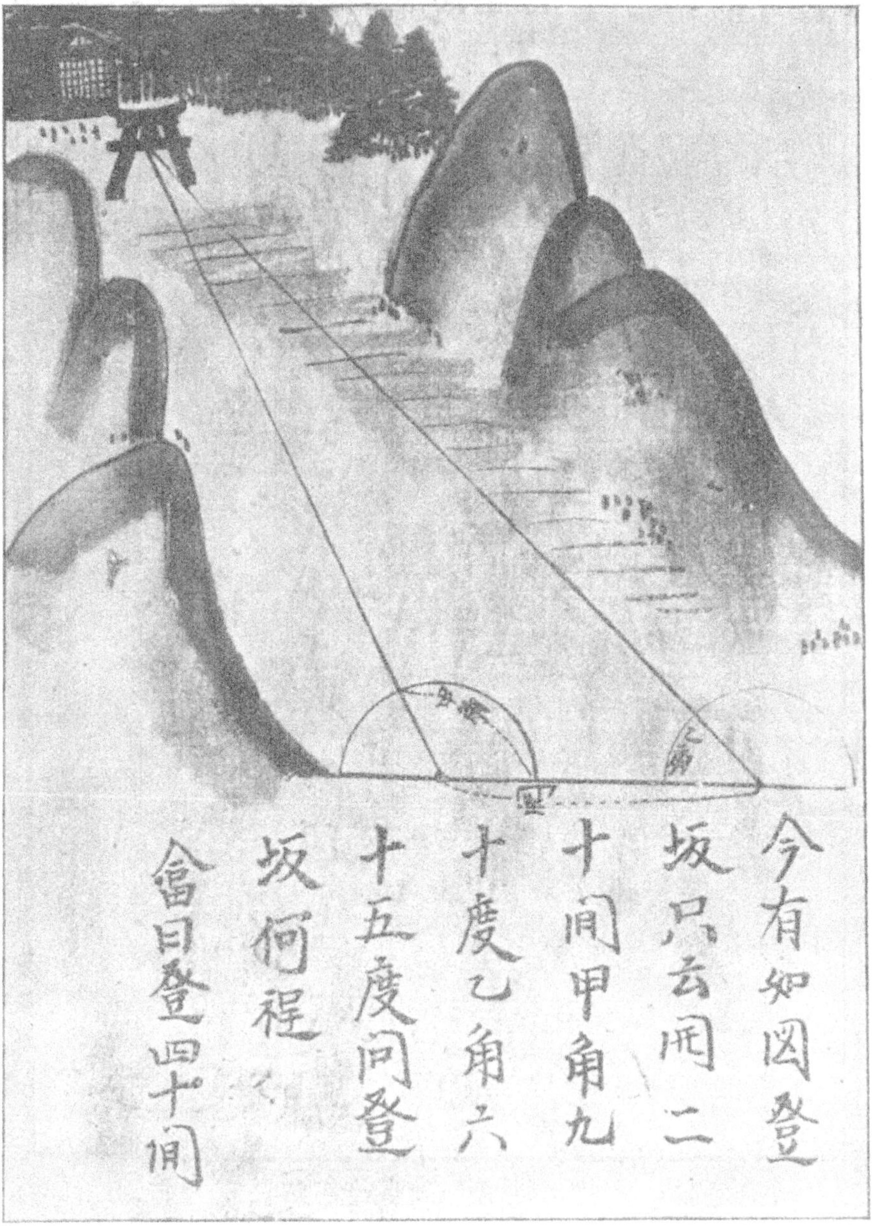

Fig. 65. From Ogino Nobutomo's *Kiku Genpō Chōken* (1718).

about the same time (Fig. 66) show distinctly the European influence.

Later writers carried the subject of trigonometry still further. For example, in Lord Arima's *Shūki Sampō* of 1769 there appear some problems in spherical trigonometry, and in Sakabe's *Sampō Tenzan Shinan-roku* of 1810—1815 the work is even more advanced. Manuscripts of Ajima and Takahashi upon the same subject are also extant. Yegawa Keishi's treatise

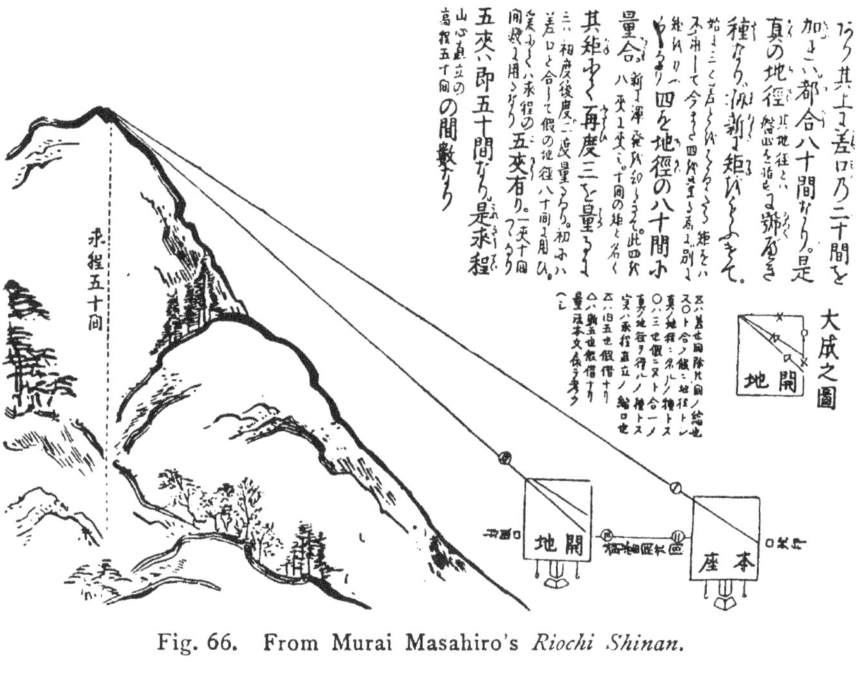

Fig. 66. From Murai Masahiro's *Riochi Shinan.*

on spherical trigonometry appeared in 1842. Some of the illustrations of the manuscripts on surveying are of interest, as is seen in the reproductions from Igarashi Atsuyoshi's *Shinki Sokurio hō* of about 1775 (Fig. 67) and from a later anonymous work (Fig. 68).

The European arithmetic began to find its way into Japan in the eighteenth century, but it never replaced the *soroban* by the paper and pencil, and there is no particular reason why it should do so. Probably the West is more likely to return to some form of mechanical calculation, as evidenced in the recent remarkable advance in calculating machinery, than is the Eastern and Russian and much of the Arabian mercantile life to give up entirely the abacus. Napier's rods, however, appealed to the Japanese and Chinese computers, and books upon their use were written in Japan. Arithmetics on the foreign plan were, however, published, Arizawa Chitei's *Chūsan Shiki* of 1725 being an example. In this work Arizawa speaks of the "Red-bearded men's arithmetic," the Japanese of

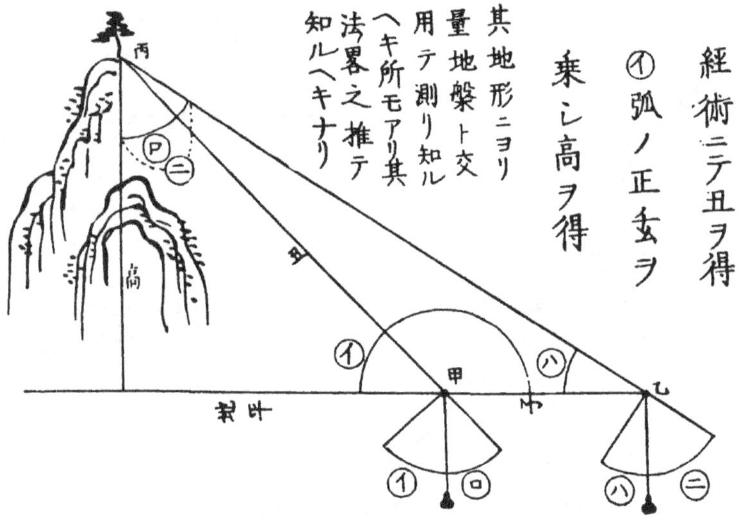

Fig. 67. From Igarashi Atsuyoshi's *Shinki Sokurio hō.*

the period sometimes calling Europeans by this name,—the title *Barbarossa* of the medieval West. Senno's works of 1767 and 1768 were upon the same subject, not to speak of several others, including Hanai Kenkichi's *Seisan Sokuchi* as late as the Ansei (1854—1860) period. (See Fig. 69.) It is a matter of tradition that Mayeno Ryōtaku (1723—1803) received an arithmetic in 1773 from some Dutch trader, but nothing is known of the work. Mayeno was a physician, and in 1769, at the age of forty-six, he began those linguistic studies that made him well known in his country. He translated several Dutch works, including a few on astronomy, but we have no

Fig. 68. From an anonymous manuscript on surveying.

evidence of his having studied European mathematics. Nevertheless one cannot be in touch with the scientific literature of a language without coming in contact with the general trend

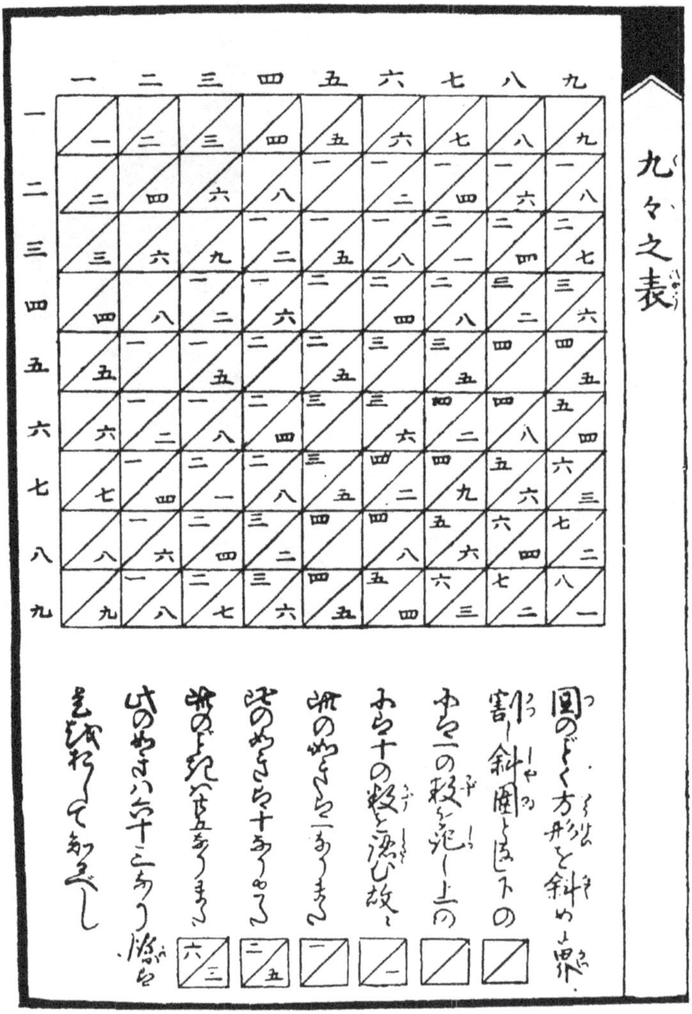

Fig. 69. From Hanai Kenkichi's *Seisan Sokuchi*, showing the Napier rods.

of thought in various lines, and it is hardly possible that Mayeno failed to communicate to mathematicians the nature of the work of their unknown *confrères* abroad.

Contemporary with Mayeno was scholar by the name of Shizuki Tadao (1760—1806),[1] an interpreter for the merchants at Nagasaki. At the close of the eighteenth century, he began a work entitled *Rekishō Shinshō,*[2] consisting of three parts, each containing two books, the composition of which was completed in 1803. The treatise, which was never printed, is based upon the works of John Keill.[3] The first part treated of the Copernican system of astronomy and the second and third parts of mechanical theories. The latter part of the work may have had its inspiration in Newton's *Principia.* It was the first Japanese work to treat of mechanics and physics, and it is noteworthy also from the fact that the appendix to the third part contains a nebular hypothesis that is claimed to have been independent of that of Kant and Laplace. Since by the statement of Shizuki his theory dated in his own mind from about 1793,[4] while Kant (1724—1804) had suggested it as early as 1755, although Laplace (1749—1827) did not publish his own speculations upon it until 1796,[5] he may have received some intimation of Kant's theory. Nevertheless Laplace is known to have stated his theory independently, so that Shizuki may reasonably be thought to have done the same.

It should also he stated that in Aida Ammei's manuscript entitled *Sampō Densho Mokuroku* (A list of Mathematical Compositions) mention is made of an *Oranda Sampō* (Dutch arithmetic). This must have been about 1790.

Contemporary with Shizuki was the astronomer Takahashi Shiji, who died in 1804, aged forty. He was familiar with the

1 He is represented in ENDŌ's *History,* Book III, p. 36, as Nakano Ryūho, Ryūho being his *nom de plume,* and the date of his book is given as 1797.

2 New Treatise on subjects relating to the theory of Calendars.

3 John Keill (1671—1721), professor of astronomy at Oxford. It is said by Dr. Korteweg to have been based upon a Dutch translation of these works, but we fail to find any save the Latin editions.

4 K. KANO, *On the Nebular Theory of Shizuki Tadao* (in Japanese), in the *Tōyō Gakugei Zasshi,* Book XII, 1895, pp. 294—300.

5 *Exposition du Système du Monde,* Paris 1796.

Dutch works upon his subject, and his writings contain extracts from some one by the name of John Lilius[1] and from various other European scholars.

The celebrated geographer Inō Chūkei (1745—1821), whose great survey of Japan has already been mentioned, was a pupil of Takahashi's, who translated La Lande, and thus came to know of the European theory of his subject, which he carried out in his field work. It might also be said that the shape of the native Japanese instruments used by surveyors early in the nineteenth century (see Fig. 70) were not unlike those in use in Europe. They were beautifully made and were as accurate as could be expected of any instrument not bearing a telescope. It should be added that Inō was not the first to use European methods in his surveys, for Nagakubo Sekisui of Mito learned the art of map drawing from a Dutchman in Nagasaki, and published a map on this plan in 1789.

Takahashi Shiji's son, Takahashi Kageyasu,[2] was also a Shogunate astronomer and as already related he died in prison for having exchanged maps with a German scientist in the Dutch service. This scientist was Philip Franz von Siebold (1796—1866), the first European scientist to explore the country. He was born at Würzburg, Germany, and attended the university there. In 1822 he entered the service of the King of the Netherlands as medical officer in the East Indian army, and was sent to Deshima, the Dutch settlement at Nagasaki. His medical skill enabled him to come in contact with Japanese people of all ranks, and in this way he had comparatively free access to the interior of the country. Well trained as a scientist and well supplied with scientific instruments and with a considerable number of native collectors, he secured a large amount of scientific information concerning a people whose

---

[1] This is recorded in the list of his writings prepared by Shibukawa Keiyū, Takahashi's second son. The name there appears in Japanese letters as Ririusu, with the usual transliteration of *r* for *l*. Very likely it was something from the writings of the well-known astrologer William Lilly.

[2] Also called Takahashi Keihō, Kageyasu being merely another reading of Keihō.

customs and country up to this time had been practically unknown to the European world. As a result he published in 1824 his *De Historia Fauna Japonica*, and in 1826 his *Epitome Linguæ Japonicæ*. He later published his *Catalogus Librorum Japonicorum, Isagoge in Bibliothecam Japonicam,* and

Fig. 70. Native Japanese surveying instrument. Early nineteenth century.

Bibliotheca Japonica, besides other works on Japan and its people. It is thus apparent that by the close of the first quarter of the nineteenth century Japan was fairly well known to the outer world, and that foreign science was influencing the work of Japanese scholars.

Indeed as early as 1811 this interrelation of knowledge had so far advanced that a Board of Translation was established in the Astronomical Observatory in Yedo, being afterward (1857) changed into an Institute for the Investigation of European Books. Both of these titles were auspicious, but they proved disappointing misnomers. Not until 1837 was any noteworthy result of the labors of the Institute apparent, and then only in the preparation of the *Seireki Shimpen* by Yamaji Kaikō[1] and a few others, and in a translation of La Lande.[2]

Foreign influence shows itself indirectly in a manuscript written in 1812 by Sakabe Kōhan. This is upon the theory of navigation and is based upon the spherical astronomy of the West. Another work along the same lines, the *Kairo Anshin-roku,* was published in 1816 by Sakabe.

In 1823 Yoshio Shunzō published his *Yensei Kanshō Zusetsu,* in three books. This work is confessedly based upon the Dutch works of Martin[3] and Martinet,[4] as is stated in the introductory note by Kusano Yōjun.[5]

---

1 Grandson of Yamaji Shujū, also a Shogunate astronomer. The work was never printed.

2 It is sometimes said that this was based on Beima's works. But Elte Martens Beima (1801—1873) wrote works on the rings of Saturn that appeared in 1842 and 1843, and there is no other Dutch writer of note on astronomy by this name.

3 Probably Martinus Martens, *Inwijings Redenvoering over eenige Voorname Nuttigheden der Wisen Sterrekunde,* Amsterdam, 1743, since Yoshio speaks of it as published sixty years earlier.

4 Johannes Florentius Martinet (1729—1792). His *Katechismus der Natuur* (1777—1779) is recorded by D. Bierens de Haan (*Bibliographie Néerlandaise,* Rome, 1883, p. 183) as having been translated into Japanese by Sammon Sammé, but with no information as to publication. Professor T. Hayashi, who has given scholarly attention to this subject, is able to find no trace of this translation. See his articles, *A list of Dutch Astronomical Works imported from Holland to Japan, How have the Japanese used the Dutch Books imported from Holland,* and *Some Dutch Books on Mathematical and Physical Sciences,* etc., in the *Nieuw Archief voor Wiskunde,* tweede reeks, zevede deel, and negende deel. Possibly the translation was merely Yoshio's work above mentioned, since its secondary title is *Catechism of Science.*

5 The work was published by him as having been orally dictated by Yoshio Shunzō.

In the Tempō Period (1830—1844) Koide Shūki translated some portions of Lalande's work on astronomy, and showed to the Astronomical Board the superiority of the European calendar, but without noticeable effect.[1]

In 1843 Iwata Seiyō published his *Kubō Shinkei Shinō* (a work relating to the telescope) in which he made use of European methods in astronomy.[2]

Fig. 71. Native Japanese surveying instrument. Early nineteenth century.

In 1851 Watanabe Ishin published a work on Illustrating the Use of the Octant, in which he even adopted the Latin term as appears by the title,—*Okutanto Yōhō Ryaku-zusetsu.* He was followed by Murata Tsunemitsu in 1853 on the use of the sextant. An octant had been brought from Europe in 1780,

[1] FUKUDA, *Sampō Tamatebako*, 1879.

[2] ENDŌ, Book III, p. 131.

but had been kept in the storehouse of the observatory because no one on the Shogunate Astronomical Board knew how to use it. Finally Yamaji Kaikō and a few others worked with it until they understood it, and Watanabe, who was an expert in gunnery, wrote the work above mentioned. He, however, was not aware of its use in astronomy, only showing how it might be employed in measuring distances in surveying.[1] The sextant was imported somewhat later than the octant, but its use was not understood until Murata Tsunemitsu published his work, and even then it was employed only in terrestrial mensuration.[2]

The Japanese first learned of logarithms through the Chinese work, the *Su-li Ching-yün*, printed at Peking in 1713. This was not the only Chinese publication of the subject, however, for it is a curious fact that no complete edition of Vlacq's tables[3] appeared in Europe after his death, and that the next publication[4] thereafter was in Peking in 1721,[5] a monument to Jesuit learning. The effect of these Chinese works was not marked, however. Ajima, who died in 1798, was one of the first Japanese mathematicians to employ logarithms in practical calculation, and his manuscript upon the subject was used by Kusaka in writing the *Fukyū Sampō* (1798), but the tables were not printed. A page from an anonymous table in an undated manuscript entitled *Tai shin Rio su hio*, giving the logarithms to seven decimal places is shown in the illustration (Fig. 72). The first printed work to suggest the actual use of the tables was Book XII of Sakabe's *Sampō Tenzan Shinan-roku* (Treatise on Tenzan Algebra), published in 1810—1815. Speaking of them he says: "Although these tables save much labor, they are but little known for the reason that they have

---

[1] ENDŌ, Book III, p. 141.

[2] ENDŌ, ibid., p. 143.

[3] His *Logarithmica Arithmetica* appeared at Gouda in 1628.

[4] They had been reprinted in part in GEORGE MILLER's *Logarithmicall Arithmetike*, London, 1631.

[5] *Magnus Canon Logarithmorum . . . Typis sinensibus in Aula Pekinensi jussu Imperatoris, 1721.*

never been printed in our country. If anyone who cares to copy them will apply to me I shall be glad to lend them to him and to give him detailed information as to their use." He gave the logarithms of the numbers 1—130 to seven decimal places, by way of illustration. He may possibly have

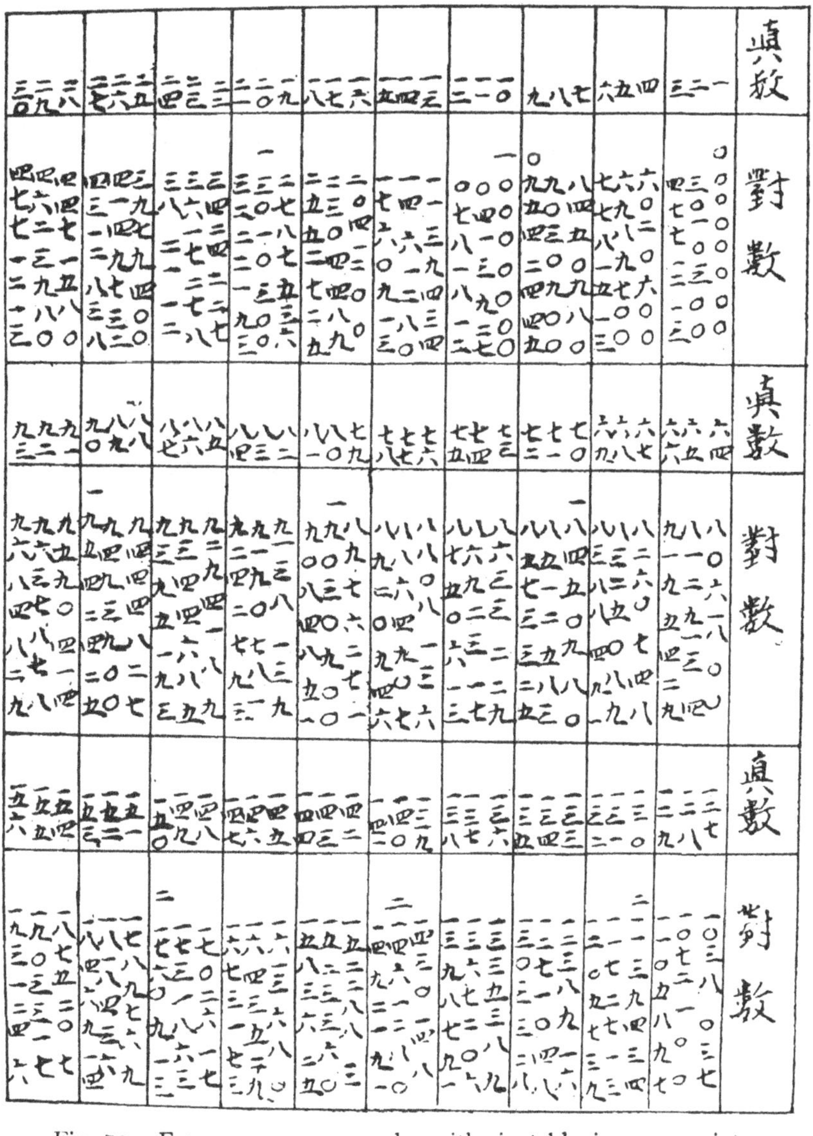

Fig. 72. From an anonymous logarithmic table in manuscript.

had some Dutch work on the subject, since he knew the word "logarithm," or possibly he had the Peking tables of 1713 and 1721.

Sakabe further says: "The ratios involved in spherical triangles, as given in the *Li-suan Ch'uan-shu,* are so numerous that it is tedious to handle them. Since addition and subtraction are easier than multiplication and division, Europeans require their calculations involving the eight trigonometric lines[1] to be made by means of adding and subtracting logarithms. They do not know, however, how to obtain the angles when the three sides are given, or how to get the sides from the three angles,[2] by the use of logarithms alone."

The first extensive logarithmic table was printed by Koide Shūki (1797—1865) in 1844. Another one was published by Yegawa Keishi in 1857, in which the logarithms were given up to 10,000,[3] and in the same year an extensive table of natural trigonometric functions was published by Okumura and Mori Masakado, in their *Katsu-yen Hio.*

Although the tables were used more or less in the first half of the nineteenth century, the theory of logarithms remained unknown for a long time after it was understood in China. Ajima, Aida, Ishiguro, and Uchida Gokan seem to have been the first to pay any attention to the nature of these numbers, but few explanations were put in print until Takemura Kōkaku published his work in 1854. Since Uchida used only logarithms to the base 10, his theory as to developing them is very complicated.[4]

It is quite probable that some suggestion leading to the study of center of gravity found its way in from the West. Seki seems the first to have had the idea in Japan, and it appears in his investigation of the volume of the solids generated by the revolution of circular arcs. Arima touches upon the subject

---

[1] I. e., the six common functions together with the versed sine and the coversed sine.

[2] Of a spherical triangle.

[3] Endō, Book III, p. 135.

[4] Endō, Book III, p. 143.

in the *Shūki Sampō* of 1769, and Takahashi Shiji also mentions it. But it was not until after the publication of Hashimoto's work in 1830, and after there was abundant opportunity for European influence to show itself, that the problem became at all popular. From that time on it was the object of a great deal of attention, the solids becoming at times quite complicated. For example, the center of gravity was studied for such a solid as a segment of an ellipsoid pierced by a cylindrical hole, and for a group of several circular cones, each piercing the others.

Similarly we may be rather sure that the study of various roulettes, including the cycloid and epicycloid, came from some hint that these problems had occupied the attention of mathematicians in the West. This does not detract from the skill shown by Wada Nei, for example, but it merely asserts that the objects, not the methods of study, were European in source. For the method, the ingenuity, and the patience, all credit is due to the Japanese scholars.

The same remark may be made with respect to the catenary and various other curves and surfaces. The catenary first appears in Hagiwara's work above mentioned, and the problem was subsequently solved by Ōmura Isshū and Kagami Mitsuteru, being attacked by approximating, step by step, the root of a transcendental equation, a treatment very complicated but full of interest. The treatment is purely Japanese, even though the idea of the problem itself may have found its way in through Dutch avenues.

In the nineteenth century there were a number of scholars in Japan who possessed more or less reading knowledge of the Dutch language. One of these was Uchida Gokan whose name has just been mentioned in connection with logarithms. He even called his school by the name "Matemateka."[1] Of him Tani Shōmo wrote, in the preface of a work published in 1840,[2] these appreciative words: "Uchida is a profoundly

---

1 Endō, Book III, p. 102.

2 Kemmochi's *Tan-i Sampō*.

learned man, and his knowledge is exceeding broad. He is master even of the 'mathematica' of the Western World." To this knowledge his sole surviving pupil, C. Kawakita, has borne witness in personal conversation with one of the authors of this history, and N. Okamoto still has some of the European books formerly owned by Uchida. Mr. Kawakita assures us, however, that Uchida's higher mathematics was his own and was not derived from Dutch sources, meaning that the method of treatment was, as we have already asserted, purely Japanese.

In a manuscript[1] written in 1824 Ichino Mokyō tells of an ellipsograph that Aida Ammei designed from a drawing in some Dutch work. "In reading some Occidental works recently," he says, "we have seen recorded a method of drawing an ellipse that is at the same time very simple and very satisfactory," and he speaks of the fact that the rectification of the ellipse by Japanese scholars is entirely original with them. Indeed it would seem that the scholars of the early nineteenth century were quite doubtful as to the superiority of the European mathematics over their own, which is a rather unexpected testimony to the independence of the Japanese in this science. Thus Oyamada Yosei uses these words upon the subject:[2] "Mogami Tokunai says in his *Sokuryō Sansaku* that the mathematical science of our country is unsurpassed by that of either China or Europe." In the same spirit an anonymous writer of the early part of the nineteenth century writes[3] these words: "There is an Occidental work wherein the value of the circumference of a circle is given to fifty figures, and of this I possess a translation which I obtained from Shibukawa. It is said that this is fully described by Montucla in his History of the Quadrature of the Circle, published in 1754,[4] but I under-

1 The *Dayen-shū Tsūjutsu* (General Method of Rectifying the Ellipse).

2 In the *Matsunoya Hikki,* an article on Mathematics and the Soroban, written early in the nineteenth century.

3 Unpublished manuscript.

4 Jean Étienne Montucla, *Histoire des recherches sur la quadrature du cercle*, Paris, 1754.

stand that this work has not been brought to Japan. I, however, have also calculated of late, with the help of Kubodera, the value to sixty figures, and not in a single figure does it differ from the European result. This goes to show how exact should be all mathematical work, and how, when this accuracy is attained, the results are the same even though the calculations be made by men who are thousands of miles apart." The same writer also says:[1] "Although the Europeans highly excel in all matters relating to astronomy and the calendar, nevertheless their mathematical theories are inferior to those that we have so accurately developed. I one time read the *Su-li Ching-yün,* compiled by Imperial edict, and in it I found a method of solving a right triangle for integral sides, . . . but the process was much too cumbersome and it was lacking in directness. . . . Moreover I have found certain problems that were incorrectly solved, although I shall not mention them specifically at this time. From this we may conclude that foreign mathematics is not on so high a plane as the mathematics of our own country."

Even such a writer as Koide Shūki had a similarly low estimate of the mathematics of the West, for he expressed himself in these words:[2] "Number dwells in the heavens and in the earth, but the arts are of human make, some being accurate and others not. The minuteness of our mathematical work far surpasses that to be found in the West, because our power is a divine inheritance, fostered by the noble and daring spirit of a nation that is exalted over the other nations of the world."

In similar spirit, the lordly spirit of the old *samurai,* Takaku Kenjirō (1821—1883) writes in his General View of Japanese Mathematics:[3] "Astronomy and the physical sciences as found in the West are truth eternal and unchangeable, and this we must learn; but as to mathematics, there Japan is leader of the world."

---

1 In his *Sanwa Zuihitsu.*

2 In his preface to KEMMOCHI's *Tan-i Sampō,* 1840.

3 For this we are indebted to the writings of C. KAWAKITA.

Hagiwara Teisuke (1828—1909), one of the last of the native school, also bemoaned the sacrifice of the *wasan* that followed on the inroads of Western science. Of his own published problems he was wont to say that no European mathematician could ever have solved them because of their very complicated nature.

Such testimony may be looked upon by some as a display of pitiful ignorance, as in certain respects it was. But on the other hand it bears testimony to the fact that the mathematicians of the old school were not looking to Europe for assistance, feeling rather that Europe should look to them.

In view of these opinions it is of interest to read the words of a serious observer of things Japanese in the seventeenth century. Engelbert Kaempfer living in Japan during the rule of the fifth of the Tokugawa Shoguns (1680—1709) remarked "They know nothing of mathematics, especially of their profound and speculative parts. No one interests himself in this science as we Europeans do."[1]

The differential and integral calculus, in its definite Western form, entered Japan through a Chinese version of the American work by Loomis.[2] This version, entitled *Tai-wei-chi Shih-chi,* was translated by Li Shan-lan in 1859, with the help of Alexander Wylie, an English missionary. About the same time several other treatises were translated into Chinese, but how many of these found their way into Japan we do not know.

As to arithmetic some mention has already been made of the European influence. Yamamoto Hokuzan says, in his preface to Ōhara Rimei's *Tenzan-Shinan* of 1810, that the tenzan algebra of the Seki school was merely founded on the European method of computing. For this statement there

---

1 KAEMPFER's work was translated from the German by SCHEUCHZER, and published in London in 1727—1728. This extract comes through a German retranslation from the English, by P. HARZER, *loc. cit.*, p. 17.

2 Elias Loomis (1811—1899). Since the work is on both algebra and the calculus it was probably compiled from the *Elements of Algebra*, New York, 1846. and the *Elements of Analytical Geometry and of the Differential and Integral Calculus,* New York, 1850.

seems to be no basis, but it shows that even in the nineteenth century the Western methods of computation were not at all well known.

About the middle of the century the European methods began to find definite place in Japanese works, if not in the

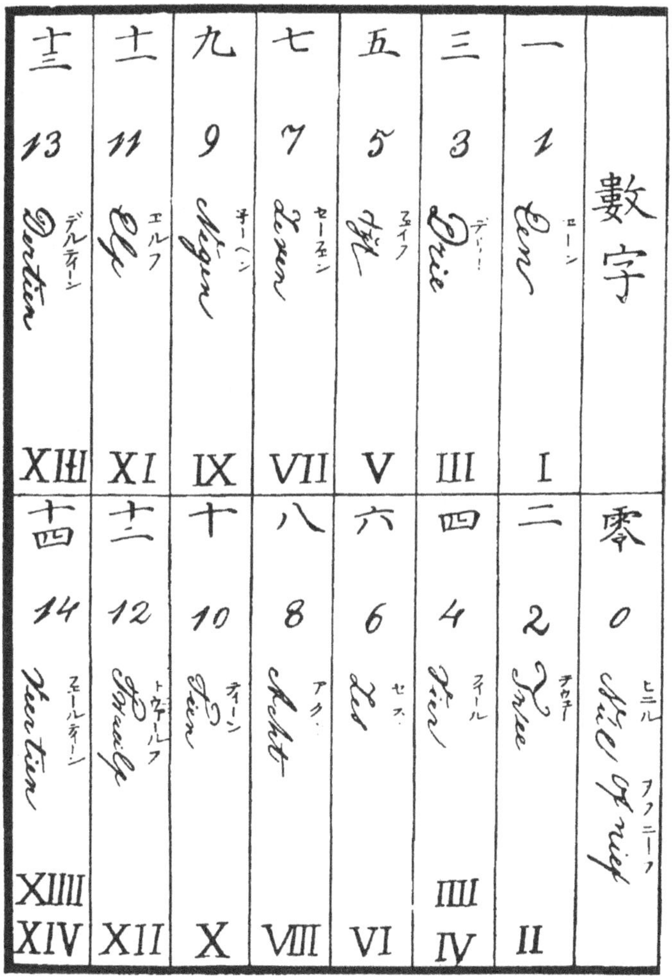

Fig. 73. From Hanai Kenkichi's *Seisan Sokuchi* (1856).

schools. The first of these works was Hanai Kenkichi's *Seisan Sokuchi* (Short Course in Western Arithmetic), published in 1856 (Fig. 73), and Yanagawa Shunzō's *Yōsan Yōhō* (Methods of Western Arithmetic) that appeared in the same year. The influence of these and similar books of later date has been on pedagogical and commercial rather than on mathematical lines. The *soroban* is as popular as ever, and save for those who proceed to higher mathematics it seems destined to remain so.

It was about the year 1851 that the Shogunate ordered certain persons to be instructed by Dutch masters at Nagasaki in the art of navigation. As a basis for this instruction Dutch arithmetic was taught and this seems to have been the first systematic instruction in the subject in Japan. In 1855 an institute was founded in Yedo for the same purpose, Dutch teachers being employed. One of the pupils in this school was Ono Tomogorō, and from him we know of the work there given.[1] The course extended over four or five years, and Li's version of the work of Loomis was used as a text-book.[2]

The influence of such a work as that of Loomis was very slight, however. Scholars who knew European mathematics were few, and the subject was generally looked upon as of inferior merit. It was not until a generation later that the Western calculus attracted much attention. Some of the efforts at combining Eastern and Western mathematics at about this period are interesting, as witness an undated manuscript by one Wake Yukimasa, of which a page is here shown (Fig. 74).

There exists in the library of one of the authors a manuscript translation from the Dutch of Jacob Floryn (1751—1818), entitled *Shinyakuho Sankaku Jutsu* (Newly translated art of trigonometry). It was made in the Ansei period (1854—1860) by Takahashtri Yoshiyasu, probably a member of the family of well-known mathematicians. It is possibly from Floryn's

---

1 *The Use of Japanese Mathematics* (in Japanese) in the *Sūgaku Hōchi*, no. 88.

2 Mr. K. UYENO informs us that the Loomis book was brought to Japan before Li's translation was made, but that there was no one who knew both English and mathematics well enough to read it.

Gröndbeginzels der Hoogere Meetkunde which was published in Rotterdam in 1794. This translation seems not to be known.

Of the conic sections some intimation of the subject may have reached Japan in the seventeenth century, but it evidently was taken, if at all, only as a hint. The Japanese studied the ellipse very zealously, always by their own peculiar

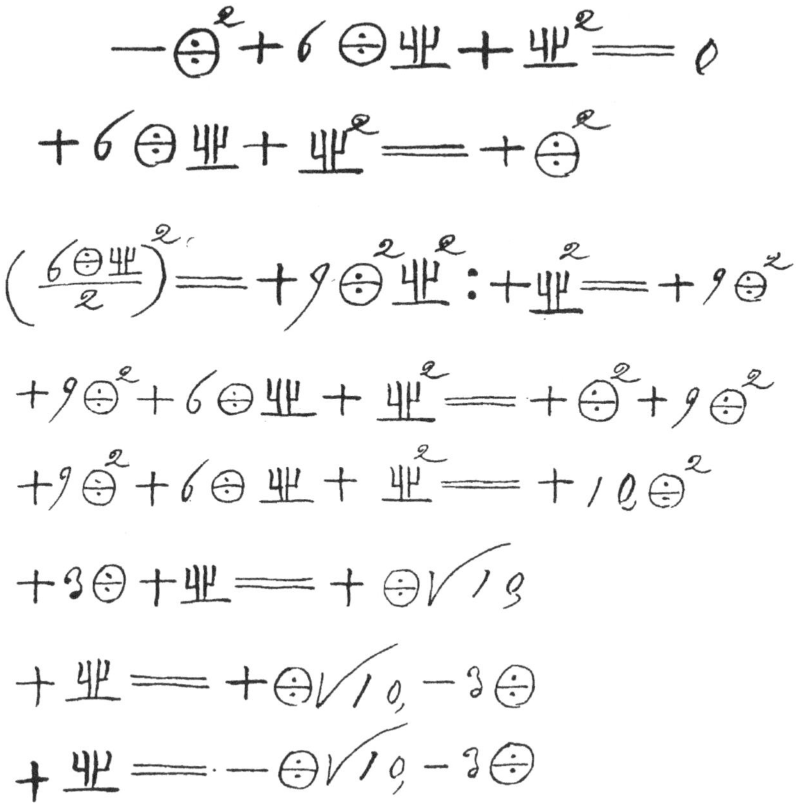

Fig. 74. From a manuscript by Wake Yukismasa.

method, but the parabola and hyperbola seem never to have attracted the attention of the old school. To be sure, the parabola enters into a problem about the path of a projectile in Yamada's *Kaisanki* of 1656, but it seems never to have been noticed by subsequent writers. The graphs of these curves are found in certain astronomical works, as in Yoshio's *Yensei Kanshō Zusetsu* of 1823 where they are used in illustrating the orbits of comets, but they do not enter into the works on pure mathematics. This very fact is evidence against any influence from without affecting the native theories.

We have already spoken of the change of the Board of Translation to the Institute for the Investigation of European Books. Six years after this change was made the Kaiseijo School was founded (1863), in which every art and science was to be taught. A department of mathematics was included, and in this Kanda Kōhei was made professor. He it was who made the first decisive step towards the teaching of European mathematics in Japan, and from his time on the subject received earnest attention in spite of the small number of students in the department.

The year 1868 is well known in the West and in Japan as a year of great import to the world. This was the year of the political revolution that overthrew the Tokugawa Shogunate, that put an end to the feudal order, and that restored the Imperial administration. Yedo, the Shogun's capital, became Tōkyō, the seat of the Empire. The year is known to the West because it marked the coming of a new World Power. What this has meant the past forty years have shown; what it is to mean as the centuries go on, no one has the slightest conception. To Japan the year marks the entrance of Western ideas, many of which are good, and many of which have been harmful. The art of Japan has suffered, in painting, in sculpture, and especially in architecture. The exquisite taste of a century ago, in textiles for example, has given place to a catering to the bad taste of moneyed tourists. And all of this has its parallel in the domain of mathematics, in which domain we may now take a retrospective view.

What of the native mathematics of Japan, and what of the effect of the new mathematics? What did Japan originate and what did she borrow? What was the status of the subject before the year 1868, and what is its status at the present and its promise for the future?

Looked at from the standpoint of the West, and weighing the evidence as carefully and as impartially as human imperfections will allow, this seems to be a fair estimate of the ancient *wasan*:—

The Japanese, beginning in the seventeenth century, produced a succession of worthy mathematicians. Since these men studied the general lines that interested European scholars of a generation earlier, and since there was some opportunity for knowing of these lines of Western interest, it seems reasonable to suppose that they had some hint of what was occupying the attention of investigators abroad. Since their methods of treatment of every subject were peculiar to Japan, either her scholars did not value or, what is quite certain, did not know the detailed methods of the West. Since they decried the European learning in mathematics, it is probable that they made no effort to know in detail what was being done by the scholars of Holland and France, of England and Germany, of Italy and Switzerland.

With such intimation as they may have had respecting the lines of research in the West, Japan developed a system of her own for the use of infinite series in the work of mensuration. She later developed an integral calculus that was sufficient for the purposes of measuring the circle, sphere, and ellipse. In the solution of higher numerical equations she improved upon the work of those Chinese scholars who had long anticipated Horner's method in England. In the study of conics her scholars paid much attention to the ellipse but none to the parabola and hyperbola.

But the mathematics of Japan was like her art, exquisite rather than grand. She never develpoed a great theory that in any way compares with the calculus as it existed when Cauchy, for example, had finished with it. When we think of Descartes's *La Géométrie*; of Desargues's *Brouillon proiect*, of the work of Newton and Leibnitz on the calculus; of that of Euler on the imaginary, for example; of Lagrange and Gauss in relation to the theory of numbers; of Galois in the discovery of groups, — and so on through a long array of names, we do not find work of this kind being done in Japan, nor have we the slightest reason for thinking that we ought to find it.

Europe had several thousand years of mathematics back of her when Newton and Leibnitz worked on the calculus,—years in which every nation knew or might know what its neighbors were doing; years in which the scholars of one country inspired those of another. Japan had had hardly a century of real opportunity in mathematics when Seki entered the field. From the standard of opportunity Japan did remarkable work; from the standpoint of mathematical discovery this work was in every way inferior to that of the West.

When, however, we come to execution it is like picking up a box of the real old red lacquer,—not the kind made for sale in our day. In execution the work was exquisite in a way wholly unknown in the West. For patience, for the everlasting taking of pains, for ingenuity in untangling minute knots and thousands of them, the problem-solving of the Japanese and the working out of some of the series in the *yenri* have never been equaled.

And what will be the result of the introduction of the new mathematics into Japan? It is altogether too early to foresee, just as we cannot foresee the effect of the introduction of new art, of new standards of living, of machinery, and of all that goes to make the New Japan. If it shall lead to the application of the peculiar genius of the old school, the genius for taking infinite pains, to large questions in mathematics, then the world may see results that will be epoch making. If on the other hand it shall lead to a contempt for the past, and to a desire to abandon the very thing that makes the *wasan* worthy of study, then we cannot see what the future may have in store. It is in the hope that the West may appreciate the peculiar genius that shows itself in the works of men like Seki, Takebe, Ajima, and Wada, and may be sympathetic with the application of that genius to the new mathematics of Japan, that this work is written.

INDEX

Featured Titles from Westphalia Press

Issues in Maritime Cyber Security Edited by Nicole K. Drumhiller, Fred S. Roberts, Joseph DiRenzo III and Fred S. Roberts

While there is literature about the maritime transportation system, and about cyber security, to date there is very little literature on this converging area. This pioneering book is beneficial to a variety of audiences looking at risk analysis, national security, cyber threats, or maritime policy.

The Rise of the Book Plate: An Exemplative of the Art by W. G. Bowdoin, Introduction by Henry Blackwel

Bookplates were made to denote ownership and hopefully steer the volume back to the rightful shelf if borrowed. They often contained highly stylized writing, drawings, coat of arms, badges or other images of interest to the owner.

The Great Indian Religions by G. T. Bettany

G. T. (George Thomas) Bettany (1850-1891) was born and educated in England, attending Gonville and Caius College in Cambridge University, studying medicine and the natural sciences. This book is his account of Brahmanism, Hinduism, Buddhism, and Zoroastrianism

Unworkable Conservatism: Small Government, Freemarkets, and Impracticality by Max J. Skidmore

Unworkable Conservatism looks at what passes these days for "conservative" principles—small government, low taxes, minimal regulation—and demonstrates that they are not feasible under modern conditions.

A Place in the Lodge: Dr. Rob Morris, Freemasonry and the Order of the Eastern Star by Nancy Stearns Theiss PhD

Ridiculed as "petticoat masonry," critics of the Order of the Eastern Star did not deter Rob Morris' goal to establish a Masonic organization that included women as members. As Rob Morris (1818-1888) came "into the light," he donned his Masonic apron and carried the ideals of Freemasonry through a despairing time of American history.

Demand the Impossible: Essays in History as Activism
Edited by Nathan Wuertenberg and William Horne

Demand the Impossible asks scholars what they can do to help solve present-day crises. The twelve essays in this volume draw inspiration from present-day activists. They examine the role of history in shaping ongoing debates over monuments, racism, clean energy, health care, poverty, and the Democratic Party.

International or Local Ownership?: Security Sector Development in Post-Independent Kosovo
by Dr. Florian Qehaja

International or Local Ownership? contributes to the debate on the concept of local ownership in post-conflict settings, and discussions on international relations, peacebuilding, security and development studies.

The Bahai Movement: A Series of Nineteen Papers
by Charles Mason Remey

Charles Mason Remey (1874-1974) was the son of Admiral George Collier Remey and grew up in Washington DC. He studied to be an architect at Cornell (1893-1896) and the Ecole des Beaux Arts in Paris (1896-1903), where he learned about the Baha'i faith, and quickly adopted it.

Ongoing Issues in Georgian Policy and Public Administration
Edited by Bonnie Stabile and Nino Ghonghadze

Thriving democracy and representative government depend upon a well functioning civil service, rich civic life and economic success. Georgia has been considered a top performer among countries in South Eastern Europe seeking to establish themselves in the post-Soviet era.

Poverty in America: Urban and Rural Inequality and Deprivation in the 21st Century
Edited by Max J. Skidmore

Poverty in America too often goes unnoticed, and disregarded. This perhaps results from America's general level of prosperity along with a fairly widespread notion that conditions inevitably are better in the USA than elsewhere. Political rhetoric frequently enforces such an erroneous notion.

---

westphaliapress.org

Made in the USA
Monee, IL
08 November 2026

67383102R00177